超低功耗单片无线系统应用入门
——基于 2.4 GHz 无线 SoC 芯片 nRF24LE1

黄智伟　杨窠江　编著

北京航空航天大学出版社

内 容 简 介

超低功耗无线收发系统应用范围广泛，2.4 GHz 超低功耗无线 SoC 芯片 nRF24LE1 是专为超低功耗无线应用设计的单片无线收发系统。本书共分 6 章，着重介绍 nRF24LE1 的主要特性、内部结构和最小系统设计，nRF24LE1 的 MCU 与应用，nRF24LE1 的接口与应用，nRF24LE1 的射频收发器与应用，nRF24LE1 与常用外围模块的连接及编程，以及 Keil μVision4 集成开发环境和 ISP 下载。本书通过大量的示例程序说明 nRF24LE1 的应用方法与技巧，所有程序都通过了验证，具有很好的工程性和实用性。本书提供所有程序源代码，读者可在北京航空航天大学出版社网站"下载专区"下载。

本书可作为电子工程技术人员进行超低功耗无线收发系统设计的参考书，也可作为高等院校本科和高职高专院校电子信息工程、通信工程、自动化、电气、计算机应用等专业学习无线收发系统设计、电子设计竞赛、课程设计、毕业设计的培训教材和教学参考书。

图书在版编目(CIP)数据

超低功耗单片无线系统应用入门：基于 2.4 GHz 无线 SoC 芯片 nRF24LE1 / 黄智伟，杨案江编著. ─ 北京：北京航空航天大学出版社，2011.7
 ISBN 978-7-5124-0425-0

Ⅰ. ①超… Ⅱ. ①黄… ②杨… Ⅲ. ①单片微型计算机②无线电通信—通信系统 Ⅳ. ①TP368.1②TN92

中国版本图书馆 CIP 数据核字(2011)第 078019 号

版权所有，侵权必究。

超低功耗单片无线系统应用入门
──基于 2.4 GHz 无线 SoC 芯片 nRF24LE1
黄智伟　杨案江　编著
责任编辑　王慕冰　王平豪　朱胜军

*

北京航空航天大学出版社出版发行

北京市海淀区学院路 37 号(邮编 100191)　http://www.buaapress.com.cn
发行部电话：(010)82317024　传真：(010)82328026
读者信箱：emsbook@gmail.com　邮购电话：(010)82316936
北京时代华都印刷有限公司印装　各地书店经销

*

开本：787×960　1/16　印张：20.75　字数：465 千字
2011 年 7 月第 1 版　2011 年 7 月第 1 次印刷　印数：4 000 册
ISBN 978-7-5124-0425-0　　定价：39.00 元

前　言

超低功耗无线收发系统可应用于无线鼠标、无线键盘、遥控游戏等 PC 机外设,以及音频/图像娱乐中心和家庭应用的遥控装置,货物追踪及监控的有源 RFID 和传感器网络,报警和访问控制的安全系统,可穿戴式传感器、运动手表,遥控智能玩具和装置等。而在这些应用中,无线收发系统的设计一直是无线应用的一个瓶颈。对于缺少无线收发系统设计经验的工程技术人员来说,单片无线发射与接收系统的出现,为解决这一难题提供了一个有效的途径。

2.4 GHz 超低功耗无线 SoC 芯片 nRF24LE1 是专为超低功耗无线应用设计的单片无线收发系统。芯片内部集成了与 8051 指令兼容的高性能 Flash 单片机、16 KB Flash 存储器、1 KB RAM、1 KB NV 非易失存储器、512 字节 NV 非易失数据存储器、实时定时器/计数器、AES 硬件加密器、16～32 位乘法/除法协处理器、随机数发生器等功能模块,以及为低功耗设计的多种电源模式,支持硬件调试;可以提供 SPI、2 线、UART、6～12 位 ADC、PWM、模拟比较器等外设接口;提供 3 种不同封装形式。nRF24LE1 具有高安全性、低功耗以及高抗干扰的优良性能,是一个较为理想的无线应用平台。

本书的特点是以应用和开发 nRF24LE 超低功耗单片无线系统所需要的知识点为基础,以示例为模板,通过大量的示例说明 nRF24LE1 的应用方法与技巧,所有示例程序都通过了验证,叙述详尽清晰,具有很好的工程性和实用性。

本书可作为电子工程技术人员进行超低功耗无线收发系统设计的参考书,也可作为高等院校本科和高职高专院校电子信息工程、通信工程、自动化、电气、计算机应用等专业学习无线收发系统设计、电子设计竞赛、课程设计、毕业设计的培训教材和教学参考书。

全书共分 6 章。各章的内容如下:

第 1 章为"超低功耗单片无线系统",介绍 nRF24LE1 的主要特性和内部结构,24 引脚、32 引脚和 48 引脚 QFN 封装的 nRF24LE1 最小系统设计,nRF24LE1 与计算机串口的连接电路设计。

第 2 章为"nRF24LE1 的 MCU 与应用",介绍 MCU 的内部结构与主要特性,存储器和 I/O 结构,MCU 的特殊功能寄存器,Flash 存储器特性、配置和编程,随机存储器(RAM)结构与功能和示例程序,定时器/计数器结构、特性、功能和示例程序,中断源、中断向量和中断用特殊功能寄存器以及示例程序,看门狗结构、功能和示例程序,器件工作模式、功耗和时钟管理功能及示例程序,电源监控系统的结构、功能和示例程序,16 MHz 晶体振荡器/RC 振荡器、

前 言

32.768 kHz 晶体振荡器/RC 振荡器，MDU（乘除法器单元）结构、功能和示例程序，加密/解密协处理器结构与功能，随机数发生器结构、功能和示例程序。

第 3 章为"nRF24LE1 的接口与应用"，介绍 GPIO 结构、功能和示例程序，SPI 结构、功能和示例程序，UART 结构、功能和示例程序，2 线接口的结构、功能和示例程序，ADC 结构、功能和示例程序，模拟比较器结构、功能和示例程序，PWM 结构、功能和示例程序。

第 4 章为"nRF24LE1 的射频收发器与应用"，介绍射频收发器内核结构与功能、工作模式、空中速率、射频通道频率、接收功率检测、PA 控制、增强型 ShockBurst、数据和控制接口。介绍射频收发器无线数据传输应用示例的系统结构，nRF24LE1 无线收发功能配置，nRF24LE1 单片机控制，液晶显示器驱动，MP3 语音模块控制等程序流程图和程序源代码。

第 5 章为"nRF24LE1 与常用外围模块的连接及编程"，介绍 nRF24LE1 与 LED 数码管和键盘模块的连接及编程，nRF24LE1 与液晶显示器模块的连接及编程，nRF24LE1 与 DAC 的连接及编程，nRF24LE1 与 DDS 的连接及编程，nRF24LE1 与超声波测距模块的连接及编程，nRF24LE1 与步进电机驱动模块的连接及编程。

第 6 章为"Keil μVision4 集成开发环境和 ISP 下载"，介绍 Keil μVision4 集成开发环境简介，工程的建立，程序的编译，HEX 文件的生成以及 ISP 下载。

本书所有示例程序都通过验证，相关程序清单可以在北京航空航天大学出版社网站"下载中心"下载。

本书在编写过程中，参考了国内外一些相关著作和资料，参考并引用了 Nordic Semiconductor 等公司提供的技术资料和应用笔记，得到了许多专家和学者的大力支持，听取了多方面的意见和建议。李富英高级工程师对本书进行了审阅，杨案江对本书中的示例进行了编程与验证，南华大学陈文光教授、王彦副教授、朱卫华副教授、李圣副教授、郝兴恒、杨必华、张翼、李军、戴焕昌、张强、税梦玲、欧科军、谭仲书、彭湃、尹晶晶、全猛、周望、黄政中和许俊杰等也为本书的编写做了大量的工作，在此一并表示衷心的感谢。

由于我们水平有限，不足之处在所难免，敬请各位读者批评斧正。有兴趣的读者，可以发送邮件到 fuzhi619@sina.com，与本书作者进行沟通；也可以发送邮件到 emsbook@gmail.com，与本书策划编辑进行交流。

黄智伟

2011 年 2 月于南华大学

目　录

第1章　超低功耗单片无线系统 ·· 1
1.1　超低功耗单片无线系统 nRF24LE1 ··· 1
1.1.1　nRF24LE1 简介 ··· 1
1.1.2　nRF24LE1 主要特性 ·· 1
1.1.3　nRF24LE1 内部结构 ·· 3
1.2　nRF24LE1 最小系统设计 ··· 5
1.2.1　24 引脚 QFN 封装的 nRF24LE1 最小系统设计 ·············· 5
1.2.2　32 引脚 QFN 封装的 nRF24LE1 最小系统设计 ·············· 8
1.2.3　48 引脚 QFN 封装的 nRF24LE1 最小系统设计 ············ 11
1.2.4　nRF24LE1 与计算机串口的连接电路 ······························ 13

第2章　nRF24LE1 的 MCU 与应用 ··· 16
2.1　MCU 内部结构与主要特性 ··· 16
2.1.1　MCU 内部结构 ··· 16
2.1.2　MCU 主要特性 ··· 17
2.2　存储器和 I/O 结构 ··· 18
2.2.1　存储器映射 ·· 18
2.2.2　PDATA 存储器寻址 ··· 18
2.2.3　MCU 特殊功能寄存器 ·· 19
2.3　Flash 存储器 ··· 24
2.3.1　Flash 存储器特性 ··· 24
2.3.2　Flash 存储器配置 ··· 25
2.3.3　MCU 对 Flash 编程 ·· 28

目 录

- 2.3.4 通过 SPI 对 Flash 编程 …… 31
- 2.3.5 硬件支持固件升级 …… 35
- 2.4 随机存储器 RAM …… 36
  - 2.4.1 随机存储器 RAM 结构与功能 …… 36
  - 2.4.2 SRAM 示例程序流程图 …… 38
  - 2.4.3 SRAM 示例程序 …… 38
- 2.5 定时器/计数器 …… 42
  - 2.5.1 定时器/计数器结构与特性 …… 42
  - 2.5.2 Timer0 和 Timer1 的功能与初始化 …… 43
  - 2.5.3 Timer2 的功能与初始化 …… 47
  - 2.5.4 定时器/计数器的特殊功能寄存器 SFR …… 48
  - 2.5.5 实时时钟 RTC …… 49
  - 2.5.6 定时器/计数器示例程序流程图 …… 49
  - 2.5.7 定时器/计数器示例程序 …… 50
- 2.6 中断 …… 54
  - 2.6.1 中断源和中断向量 …… 54
  - 2.6.2 中断用特殊功能寄存器 SFR …… 55
  - 2.6.3 中断示例外接电路 …… 55
  - 2.6.4 中断示例程序流程图 …… 56
  - 2.6.5 中断示例程序 …… 56
- 2.7 看门狗 …… 61
  - 2.7.1 看门狗结构与功能 …… 61
  - 2.7.2 看门狗寄存器 WDSV …… 62
  - 2.7.3 看门狗示例程序流程图 …… 62
  - 2.7.4 看门狗示例程序 …… 62
- 2.8 功耗和时钟管理 …… 68
  - 2.8.1 工作模式 …… 68
  - 2.8.2 功耗和时钟管理有关的寄存器 …… 69
  - 2.8.3 功耗和时钟管理示例程序 …… 70
- 2.9 电源监控 …… 79
  - 2.9.1 电源监控结构与功能 …… 79
  - 2.9.2 电源监控示例程序流程图 …… 82
  - 2.9.3 电源监控示例程序 …… 82
- 2.10 片上振荡器 …… 86

2.10.1　16 MHz 晶体振荡器 …………………………………………………………… 86
 2.10.2　16 MHz RC 振荡器 ……………………………………………………………… 87
 2.10.3　外部 16 MHz 时钟 ……………………………………………………………… 87
 2.10.4　32.768 kHz 晶体振荡器 ………………………………………………………… 87
 2.10.5　32.768 kHz RC 振荡器 ………………………………………………………… 88
 2.10.6　合成 32.768 kHz 时钟 ………………………………………………………… 88
 2.10.7　外部 32.768 kHz 时钟 ………………………………………………………… 88
2.11　乘除法器单元 MDU ……………………………………………………………………… 88
 2.11.1　MDU 结构与功能 ……………………………………………………………… 88
 2.11.2　MDU 操作步骤 ………………………………………………………………… 89
 2.11.3　MDU 示例程序流程图 ………………………………………………………… 91
 2.11.4　MDU 示例程序 ………………………………………………………………… 91
2.12　加密/解密协处理器 ……………………………………………………………………… 99
2.13　随机数发生器 …………………………………………………………………………… 99
 2.13.1　随机数发生器结构与功能 ……………………………………………………… 99
 2.13.2　随机数发生器示例程序流程图 ………………………………………………… 100
 2.13.3　随机数发生器示例程序 ………………………………………………………… 100

第 3 章　nRF24LE1 的接口与应用 ………………………………………………………… 105

3.1　通用 I/O 端口 GPIO ……………………………………………………………………… 105
 3.1.1　GPIO 结构与功能 ……………………………………………………………… 105
 3.1.2　I/O 端口可编程寄存器 ………………………………………………………… 107
 3.1.3　GPIO 与按键和 LED 的连接电路 ……………………………………………… 113
 3.1.4　GPIO 示例程序流程图 ………………………………………………………… 114
 3.1.5　GPIO 示例程序 ………………………………………………………………… 114
3.2　串行外设接口 SPI ………………………………………………………………………… 117
 3.2.1　SPI 结构与功能 ………………………………………………………………… 117
 3.2.2　SPI 主模式寄存器 ……………………………………………………………… 117
 3.2.3　SPI 从模式寄存器 ……………………………………………………………… 119
 3.2.4　SPI 时序 ………………………………………………………………………… 121
 3.2.5　SPI 主设与 SPI 从设之间的互联 ……………………………………………… 123
 3.2.6　SPI 示例程序流程图 …………………………………………………………… 123
 3.2.7　SPI 示例程序 …………………………………………………………………… 124
3.3　UART ……………………………………………………………………………………… 131

目 录

- 3.3.1 UART 结构与功能 …………………………………………………… 131
- 3.3.2 UART 可编程寄存器 …………………………………………………… 132
- 3.3.3 UART 示例程序流程图 ………………………………………………… 133
- 3.3.4 UART 示例程序 ………………………………………………………… 134

3.4 2 线接口 …………………………………………………………………… 138
- 3.4.1 2 线接口结构与功能 …………………………………………………… 138
- 3.4.2 2 线接口主设发送/接收 ………………………………………………… 138
- 3.4.3 2 线接口从设发送/接收 ………………………………………………… 139
- 3.4.4 2 线接口时序 …………………………………………………………… 139
- 3.4.5 2 线接口特殊功能寄存器 ……………………………………………… 140
- 3.4.6 2 线接口应用示例电路 ………………………………………………… 143
- 3.4.7 2 线接口应用示例程序流程图 ………………………………………… 144
- 3.4.8 2 线接口应用示例程序 ………………………………………………… 144

3.5 ADC ……………………………………………………………………… 158
- 3.5.1 ADC 特性与结构 ……………………………………………………… 158
- 3.5.2 ADC 功能说明 ………………………………………………………… 159
- 3.5.3 ADC 特殊功能寄存器 ………………………………………………… 161
- 3.5.4 ADC 模拟电压输入电路 ……………………………………………… 164
- 3.5.5 ADC 示例程序流程图 ………………………………………………… 164
- 3.5.6 ADC 示例程序 ………………………………………………………… 165

3.6 模拟比较器 ……………………………………………………………… 169
- 3.6.1 模拟比较器特性与结构 ……………………………………………… 169
- 3.6.2 模拟比较器功能 ……………………………………………………… 169
- 3.6.3 模拟比较器特殊功能寄存器 ………………………………………… 170
- 3.6.4 模拟比较器示例程序流程图 ………………………………………… 171
- 3.6.5 模拟比较器示例程序 ………………………………………………… 171

3.7 PWM ……………………………………………………………………… 178
- 3.7.1 PWM 结构与功能 ……………………………………………………… 178
- 3.7.2 PWM 特殊功能寄存器 ………………………………………………… 178
- 3.7.3 电机控制和驱动电路 ………………………………………………… 180
- 3.7.4 PWM 示例程序流程图 ………………………………………………… 181
- 3.7.5 PWM 示例程序 ………………………………………………………… 181

目录

第 4 章　nRF24LE1 的射频收发器与应用 ………………………………… 185

- 4.1　nRF24LE1 的射频收发器 ……………………………………………… 185
  - 4.1.1　射频收发器内核结构与功能 ……………………………………… 185
  - 4.1.2　射频收发器工作模式 ……………………………………………… 186
  - 4.1.3　射频收发器空中速率 ……………………………………………… 189
  - 4.1.4　射频收发器射频通道频率 ………………………………………… 190
  - 4.1.5　接收功率检测 ……………………………………………………… 190
  - 4.1.6　PA 控制 ……………………………………………………………… 190
  - 4.1.7　增强型 ShockBurst ………………………………………………… 191
  - 4.1.8　数据和控制接口 …………………………………………………… 195
- 4.2　射频收发器应用示例 1 ………………………………………………… 199
  - 4.2.1　无线传输结构形式 ………………………………………………… 199
  - 4.2.2　无线传输示例程序流程图 ………………………………………… 200
  - 4.2.3　无线传输示例程序 ………………………………………………… 200
- 4.3　射频收发器应用示例 2 ………………………………………………… 224
  - 4.3.1　系统结构 …………………………………………………………… 224
  - 4.3.2　发送端电路 ………………………………………………………… 224
  - 4.3.3　接收端电路 ………………………………………………………… 225
  - 4.3.4　无线遥控 MP3 播放器示例程序流程图 …………………………… 228
  - 4.3.5　无线遥控 MP3 播放器示例程序 …………………………………… 228

第 5 章　nRF24LE1 与常用外围模块的连接及编程 …………………… 263

- 5.1　nRF24LE1 与数码管和键盘的连接及编程 …………………………… 263
  - 5.1.1　nRF24LE1 与 ZLG7289 的连接 …………………………………… 263
  - 5.1.2　nRF24LE1 与 ZLG7289 的编程示例 ……………………………… 263
- 5.2　nRF24LE1 与液晶显示器模块的连接及编程 ………………………… 273
  - 5.2.1　RT12864M 汉字图形点阵液晶显示器模块简介 ………………… 273
  - 5.2.2　nRF24LE1 与 RT12864M 的连接 ………………………………… 274
  - 5.2.3　nRF24LE1 与液晶显示器模块的编程示例 ……………………… 275
- 5.3　nRF24LE1 与 DAC 的连接及编程 …………………………………… 279
  - 5.3.1　nRF24LE1 与 DAC TLC5615 的连接 …………………………… 279
  - 5.3.2　nRF24LE1 与 DAC 的编程示例 ………………………………… 280
- 5.4　nRF24LE1 与 DDS 的连接及编程 …………………………………… 285

目 录

  5.4.1 nRF24LE1 与 DDS AD9850 的连接 …………………………………… 285
  5.4.2 nRF24LE1 与 DDS 的编程示例 ……………………………………… 286
 5.5 nRF24LE1 与超声波模块的连接及编程 …………………………………… 291
  5.5.1 nRF24LE1 与超声波模块的连接 ……………………………………… 291
  5.5.2 nRF24LE1 与超声波模块的编程示例 ………………………………… 292
 5.6 nRF24LE1 与步进电机驱动模块的连接及编程 …………………………… 301
  5.6.1 nRF24LE1 与步进电机驱动模块的连接 ……………………………… 301
  5.6.2 nRF24LE1 与步进电机驱动模块的编程示例 ………………………… 303

第 6 章 Keil μVision4 集成开发环境和 ISP 下载 ……………………………… 310

 6.1 Keil μVision4 集成开发环境的使用 ………………………………………… 310
  6.1.1 工程的建立 ……………………………………………………………… 310
  6.1.2 添加 C 语言文件 ………………………………………………………… 314
  6.1.3 代码编辑 ………………………………………………………………… 316
  6.1.4 工程编译 ………………………………………………………………… 316
  6.1.5 生成 HEX 文件 …………………………………………………………… 316
 6.2 ISP 下载 ……………………………………………………………………… 318

参考文献 ……………………………………………………………………………… 320

第 1 章
超低功耗单片无线系统

1.1 超低功耗单片无线系统 nRF24LE1

1.1.1 nRF24LE1 简介

nRF24LE1 是低成本、高性能、内嵌入微处理器的射频收发器 nRF24 系列的家族成员之一。nRF24LE1 是为超低功耗无线应用设计的单片无线收发系统,芯片内部集成了高性能微处理器(与 8051 指令兼容)、16 KB Flash 存储器、1 KB 数据空间(片内 RAM)、1 KB NV 非易失存储器空间、512 字节 NV 非易失数据存储(扩展寿命)、低功耗振荡器、实时计数器、AES 硬件加密器、16~32 位乘法/除法协处理器(MDU)、随机数发生器等功能模块,以及为低功耗设计的多种电源模式,支持硬件调试,硬件支持固件更新。nRF24LE1 提供了一个理想的无线协议平台,具有协议的无缝连接、高安全性、低功耗以及高抗干扰的优良性能。nRF24LE1 提供了 SPI、2 线接口、UART、6~12 位 ADC、PWM 和一个低功耗的可作为系统电平唤醒的模拟比较器等外设接口,提供了 3 种不同封装形式:

- 具有 7 个通用 I/O 的超小型 4 mm×4 mm 24 引脚 QFN 封装;
- 具有 15 个通用 I/O 的紧凑型 5 mm×5 mm 32 引脚 QFN 封装;
- 具有 31 个通用 I/O 的 7 mm×7 mm 48 引脚 QFN 封装。

nRF24LE1 可以满足如无线鼠标、无线键盘、遥控游戏等 PC 外设,音频/图像娱乐中心和家庭应用的先进遥控装置,货物追踪和监控的有源 RFID 和传感器网络,报警和访问控制的安全系统,可穿戴式传感器、运动手表、遥控智能玩具和装置等的应用要求。

1.1.2 nRF24LE1 主要特性

nRF24LE1 的主要特性如下:

1. 高速 8 位处理器

芯片内部嵌入高速 8 位处理器,与 Intel 8051 指令兼容。它采用简化的指令周期,与传统

第1章　超低功耗单片无线系统

的 8051 相比，速度快 12 倍，具有 32 位乘除法单元。

2．存储器

- 程序存储空间——带加密功能的 16 KB Flash 存储器（最少可擦/写 1 000 次）；
- 数据存储器——1 KB 片内 RAM；
- 非易失数据存储器——1 KB；
- 可擦/写的非易失数据存储器——512 字节（最少可擦/写 20 000 次）。

3．多功能 I/O 引脚端

具有 7~31 个多功能 I/O 引脚端（取决于不同的封装），许多片内的硬件资源可通过可编程的多功能 I/O 引脚端来实现，如 GPIO、SPI 主设、SPI 从设、主/从 2 线接口、全双工串行接口、PWM、ADC、模拟比较器、外部中断、定时器中断输入、32.768 kHz 晶体振荡器、调试接口。输入高电平为 V_{DD}，输入低电平为 $0.3V_{DD}$。输出高电平为 V_{DD}，输出低电平为 0.3 V。上拉/下拉电阻为 11~16 kΩ。

4．2.4 GHz GFSK 射频收发器

具有单片高性能 2.4 GHz GFSK 射频收发器，接收灵敏度为 −82~−94 dBm，发射功率为 0~4 dBm，工作频率范围为 2.400~2.525 GHz，天线负载阻抗为 (15+j88)Ω。硬件支持 OSI 的链路层和增强型 ShockBurst 链路层，包的装配/拆解，地址和 CRC 计算，自动 ACK（应答）和重发，空中数据速率为 250 kbps、1 Mbps 和 2 Mbps，数据接口（SPI）速率为 0~8 Mbps，具有 125 个无线频道，其中 79 个频道（2.402~2.81 GHz）在 2.400~2.4853 GHz 内，适合跳频应用的快速切换时间，无线收发特性与 nRF24Lxx 系列芯片完全兼容，兼容 nRF2401A、nRF2402、nRF24E1 和 nRF24E2 的 250 kbps 和 1 Mbps 模式。

5．ADC

ADC 具有 6 位、8 位、10 位和 12 位分辨率，14 路输入通道，单端或差分输入，可由内部基准、外部基准或 V_{DD} 设定满量程范围，单步模式的转换时间低于 3 μs，连续模式具有 2 kbps、4 kbps、8 kbps 和 16 kbps 的采样速率，低电流消耗（在 2 ksps 采样时，电流仅为 0.1 mA），具有测量电源电压模式。

6．模拟比较器

模拟比较器可差分或单端输入，具有轨到轨输入范围，14 路输入的多路复用器。其单端的阈值电压可编程（可编程到 V_{DD} 的 25%、50%、75% 和 100%，或来自引脚端的外部基准电源电压），输出极性可编程，可以作为唤醒源使用，电流消耗也极低（典型值为 0.75 μA）。

7．加密/解密器

加密/解密器具有优化的 AES 加密时间和功耗。

8. 随机数发生器

随机数发生器采用基于热噪声的非确定性结构，不需要种子值，非重复序列，校正算法确保统一的统计分布，数据速率达到 10 KBps，处理器在待机模式时仍可工作。

9. 系统复位和电源监控

具有片内上电复位和掉电、看门狗定时器、外部引脚复位功能。具有阈值可编程的电源失效比较器，并可产生中断到 MCU。

10. 片上定时器

具有在系统时钟下工作的 3 个 16 位定时器/计数器（源于片上 16 MHz 振荡器）和一个在低频时钟（32.768 kHz）下工作的 16 位定时器/计数器。

11. 片上振荡器

具有 16 MHz 晶体振荡器（XOSC16M）、16 MHz RC 振荡器（RCOSC16M）、32.768 kHz 晶体振荡器（XOSC32K）、32.768 kHz RC 振荡器（RCOSC32K）。

12. 电源管理功能

工作电压为 1.9～3.3 V，供电上升时间（0～1.9 V）最大值为 50 ms，电流消耗最大值为 13.3 mA。电源管理功能的低功耗设计支持全静态停机/待机，可编程的 MCU 时钟频率范围为 125 kHz～16 MHz，片上稳压器支持低功耗模式，看门狗和唤醒功能可以工作在低功耗模式式。工作温度范围为 −40～+85 ℃。

13. 调试工具

片上支持 FS2 或 nRFprobe 硬件调试工具，支持 Keil 开发工具。

14. 固件平台

可提供完全的固件平台，如硬件抽象层（HAL）函数、nRF24L01+ 库函数、AES HAL 以及应用实例。

1.1.3　nRF24LE1 内部结构

nRF24LE1 内部结构方框图如图 1.1.1 所示，主要由微控制器电路和射频收发器电路组成。

第1章 超低功耗单片无线系统

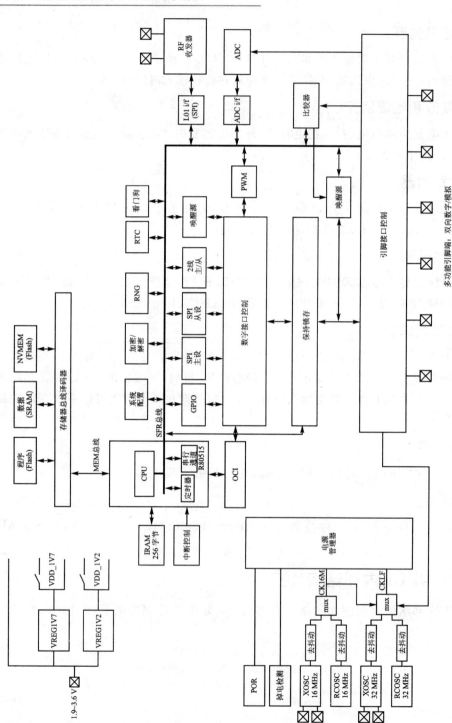

图 1.1.1 nRF24LE1 内部结构方框图

1.2 nRF24LE1 最小系统设计

1.2.1 24 引脚 QFN 封装的 nRF24LE1 最小系统设计

采用 4 mm×4 mm 24 引脚 QFN 封装的 nRF24LE1 引脚封装形式如图 1.2.1 所示,引脚端功能如表 1.2.1 所列,其中引脚端 P0.0~P0.6 具有复用功能,P0.0~P0.6 的复用功能如表 1.2.2 所列。

注意:SMISO 引脚端仅在 SCSN 引脚端有效时被使能。

厂商推荐的 24 引脚 QFN 封装的 nRF24LE1 最小系统应用电路元器件布局和 PCB 图如图 1.2.2 和图 1.2.3 所示,电路元器件清单如表 1.2.3 所列。

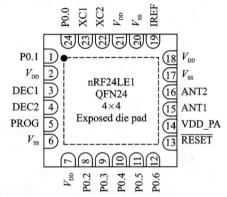

图 1.2.1 4 mm×4 mm 24 引脚 QFN 封装的 nRF24LE1 引脚封装形式

表 1.2.1 nRF24LE1 的引脚端功能

引脚端符号	类 型	功 能
V_{DD}	电源正端	+1.9~+3.6 V DC 输入
V_{SS}	电源负端	接地(0 V)
DEC1、DEC2	输出	电源退耦输出,连接 100 nF 电容器到 DEC1,连接 33 nF 电容器到 DEC2
P0.0~P3.6	I/O	数字或模拟 I/O,数量与封装形式有关
PROG	数字输入	使能 Flash 编程
$\overline{RESET}$	数字输入	微控制器复位,低电平有效
IREF	模拟输入	器件的基准电流输出,连接基准电阻到 PCB
VDD_PA	电源输出	RF 功率放大器电源输出(+1.8 V)
ANT1、ANT2	RF 天线接口	RF 差分天线连接(TX 和 RX)
XC1、XC2	模拟输入	连接 16 MHz 晶振
Exposed die pad	裸露的焊盘,接地与散热	对于 QFN48 7 mm×7 mm 和 QFN32 5 mm×5 mm 封装的器件,连接该焊盘到 GND;对于 QFN24 4 mm×4 mm 封装的器件,不能够连接到地

表 1.2.2 引脚端 P0.0～P0.6 的复用功能

引脚	默认连接		动态使能连接						
	输入	输出	XOSC32K	SPI 主设	从设/Flash SPI	HW 调试	2 线	PWM	ADC/比较器
			优先级 1	优先级 2	优先级 3	优先级 4	优先级 5	优先级 6	优先级 7
P0.6	p0Di6 / UART/RXD	p0Do6				OCITO 输出	WS2DA 输入/输出	PWM1 输出	AIN6 模拟
P0.5	p0Di5	p0Do5 / UART/TXD			SCSN 输入	OCITDO 输出	W2SCL 输入/输出		AIN5 模拟
					FCSN① 输入				
P0.4	p0Di4 / TIMER0	p0Do4		MMISO 输入	SMISO 输出	OCITDI 输入			AIN4 模拟
					FMISO① 输出				
P0.3	p0Di3	p0Do3		MMOSI 输出	SMOSI 输入	OCITMS 输入		PWM0 输出	AIN3 模拟
					FMOSI① 输入				
P0.2	p0Di2 / GPINT1	p0Do2		MSCK 输出	SSCK 输入	OCITCK 输入			AIN2 模拟
					FSCK① 输入				
P0.1	p0Di1	P0Do1	CKLP②						AIN1 模拟
P0.0	p0Di0 / GPINT0	P0Do0	CKLF③ 模拟						AIN0 模拟

① PROG 设置为高电平,仅 Flash SPI 接口有效,在运行操作时没有冲突。
② 连接取决于配置寄存器 CKLFCTL[2:0]
    CKLFCTL[2:0]=3'b000:晶振连接在引脚端 P0.0 和 P0.1 之间。
    CKLFCTL[2:0]=3'b011:低幅度的模拟时钟源连接到引脚端 P0.1。
    CKLFCTL[2:0]=3'b100:数字源。
③ 连接取决于配置寄存器 CKLFCTL[2:0]
    CKLFCTL[2:0]=3'b000:晶振连接在引脚端 P0.0 和 P0.1 之间。

第 1 章 超低功耗单片无线系统

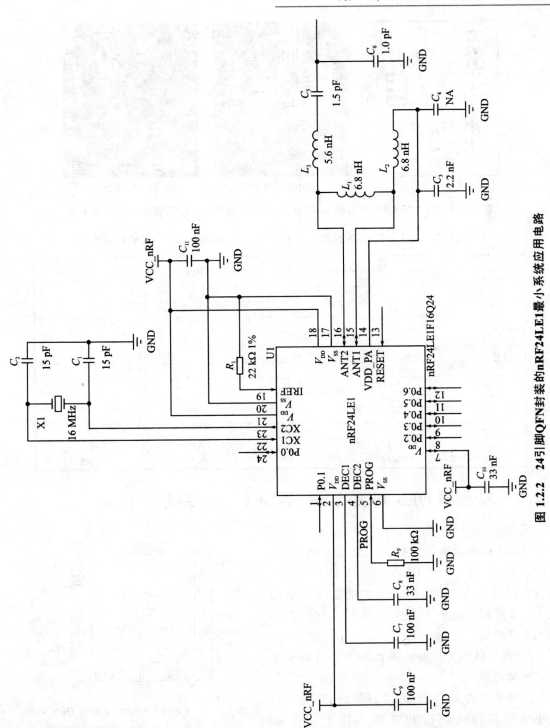

图 1.2.2 24 引脚 QFN 封装的 nRF24LE1 最小系统应用电路

第1章 超低功耗单片无线系统

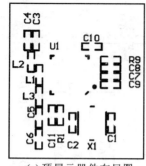 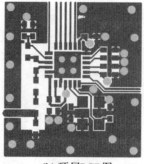 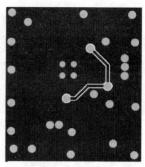

(a) 顶层元器件布局图　　(b) 顶层PCB图　　(c) 底层PCB图

图 1.2.3　24 引脚 QFN 封装的 nRF24LE1 最小系统应用电路元器件布局和 PCB 图

表 1.2.3　24 引脚 QFN 封装的 nRF24LE1 最小系统应用电路元器件清单

符　号	数　值	封装形式	备　注
C_1,C_2	15 pF	0402s	NP0　50 V
C_3	2.2 nF	0402s	X7R　16 V
C_4	NA	0402s	NP0　50 V
C_5	1.5 pF	0402s	NP0　±0.1 pF,50 V
C_6	1.0 pF	0402s	NP0　±0.1 pF,50 V
C_7,C_9,C_{11}	100 nF	0402s	X7R　16 V
C_8,C_{10}	33 nF	0402s	X7R　16 V
L_1,L_2	6.8 nH	0402s	片式电感±5%
L_3	5.6 nH	0402s	片式电感±5%
R_1	22 kΩ	0402s	±1%
R_9	100 kΩ	0402s	±1%
U1	nRF24LE1F16 Q24	QFN24	QFN24　4 mm×4 mm 封装
X1	16 MHz	—	TSX-3225,16 MHz,C_1=9 pF,±10%,10^{-6}

1.2.2　32 引脚 QFN 封装的 nRF24LE1 最小系统设计

采用 5 mm×5 mm 32 引脚 QFN 封装的 nRF24LE1 引脚封装形式如图 1.2.4 所示,其中引脚端 P0.0~P0.7 和 P1.0~P1.6 具有复用功能,P0.0~P0.7 和 P1.0~P1.6 的复用功能如表 1.2.4 所列。表 1.2.4 中的注释①、②、③与表 1.2.2 相同。

注意: SMISO 引脚端仅在 SCSN 引脚端有效时被使能。

厂商推荐的 32 引脚 QFN 封装的 nRF24LE1 最小系统应用电路和 PCB 图如图 1.2.5 和

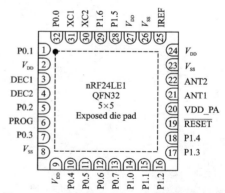

图 1.2.4　5 mm×5 mm 32 引脚 QFN 封装的 nRF24LE1 引脚封装形式

图 1.2.6 所示,电路元器件清单如表 1.2.3 所列。

注意:元器件表中 L_3 修改为 4.7 nH,U1 修改为 nRF24LE1F16 Q32。

表 1.2.4 引脚端 P0.0～P0.7 和 P1.0～P1.6 的复用功能

引脚	默认连接		动态使能配置						
	输入	输出	XOSC32K	SPI 主设	从设/Flash SPI	PWM	ADC/COMP	HW 调试	2线接口
			优先级 1	优先级 2	优先级 3	优先级 4	优先级 5	优先级 6	优先级 7
P1.6	p1Di6	p1Do6		MMISO	输入				
P1.5	p1Di5	p1Do5		MMOSI	输出				
P1.4	p1Di4	p1Do4		MSCK	输出				
P1.3	p1Di3	p1Do3						OCITO	输出
P1.2	p1Di2	p1Do2					AIN10	模拟 OCITDO	输出
P1.1	p1Di1	p1Do1			SCSN 输入		AIN9	模拟 OCITDI	输入
					FCSN① 输入				
P1.0	p1Di0 / TIMER1	p1Do0			SMISO 输出		AIN8	模拟 OCITMS	输入
					FMISO① 输出				
P0.7	p0Di7 / TIMER0	p0Do7			SMOSI 输入		AIN7	模拟 OCITCK	输入
					FMOSI① 输入				
P0.6	p0Di6 / GPINT1	p0Do6					AIN6	模拟	
P0.5	p0Di5 / GPINT0	p0Do5			SSCK 输入		AIN5	模拟	W2SDA 输入/输出
					FSCK① 输入				
P0.4	p0Di4 / UART/RXD	p0Do4					AIN4	模拟	W2SCL 输入/输出
P0.3	p0Di3	p0Do3 / UART/TXD				PWM1 输出	AIN3	模拟	
P0.2	p0Di2	p0Do2				PWM0 输出	AIN2	模拟	
P0.1	p0Di1	p0Do1	CKLF②				AIN1	模拟	
P0.0	p0Di0	p0Do0	CKLF③ 模拟				AIN0	模拟	

第 1 章　超低功耗单片无线系统

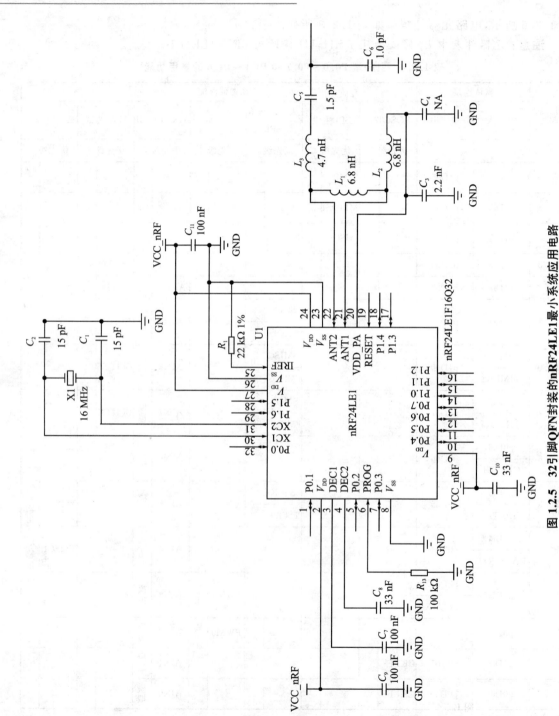

图 1.2.5　32引脚QFN封装的nRF24LE1最小系统应用电路

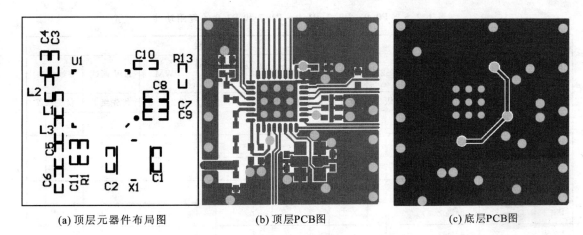

(a) 顶层元器件布局图　　(b) 顶层PCB图　　(c) 底层PCB图

图1.2.6　32引脚QFN封装的nRF24LE1最小系统应用电路PCB图

1.2.3　48引脚QFN封装的nRF24LE1最小系统设计

采用7 mm×7 mm 48引脚QFN封装的nRF24LE1引脚封装形式如图1.2.7所示,其中引脚端P0.0～P0.7、P1.0～P1.7和P2.0具有复用功能,P0.0～P0.7、P1.0～P1.7和P2.0的复用功能如表1.2.5所列,表1.2.5中的注释①、②、③与表1.2.2相同。注意,SMISO引脚端仅在SCSN引脚端有效时被使能。

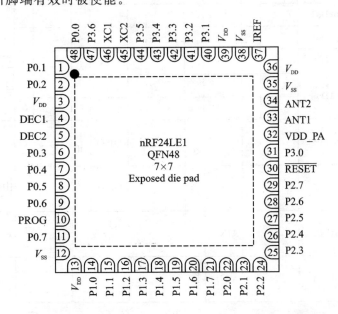

图1.2.7　7 mm×7 mm 48引脚QFN封装的nRF24LE1引脚封装形式

表 1.2.5　P0.0~P0.7、P1.0~P1.7 和 P2.0 的复用功能

引脚	默认连接		动态使能连接						
	输入	输出	XOSC32K	ADC/COMP	SPI 主设	从设/Flash SPI	PWM	HW 调试	2线接口
			优先级 1	优先级 4	优先级 2		优先级 6	优先级 5	优先级 7
P3.6	p3Di6	p3Do6							
P3.5	p3Di5	p3Do5							
P3.4	p3Di4	p3Do4							
P3.3	p3Di3	p3Do3							
P3.2	p3Di2	p3Do2							
P3.1	p3Di1	p3Do1							
P3.0	p3Di0	p3Do0							
P2.7	p2Di7	p2Do7							
P2.6	p2Di6	p2Do6							
P2.5	p2Di5	p2Do5							
P2.4	p2Di4	p2Do4							
P2.3	p2Di3	p2Do3							
P2.2	p2Di2	p2Do2							
P2.1	p2Di1	p2Do1							
P2.0	p2Di0	p2Do0				FCSN① 输入			
P1.7	p1Di7 / TIMER2	p1Do7							
P1.6	p1Di6 / TIMER1	p1Do6				FMISO① 输出			
P1.5	p1Di5 / TIMER0	p1Do5		AIN13 模拟		FMOSI① 输入		OCITO 输出	
P1.4	p1Di4 / GPINT2	p1Do4		AIN12 模拟				OCITDO 输出	
P1.3	p1Di3 / GPINT1	p1Do3		AIN11 模拟				OCITDI 输入	W2SDA 输入/输出

续表 1.2.5

引脚	默认连接		动态使能连接						
	输入	输出	XOSC32K	ADC/COMP	SPI 主设	从设/Flash SPI	PWM	HW 调试	2线接口
			优先级 1	优先级 4	优先级 2		优先级 6	优先级 5	优先级 7
P1.2	p1Di2 / GPINT0	p1Do2		AIN10	模拟	FSCK① / 输入		OCITMS / 输入	W2SCL / 输入/输出
P1.1	p1Di1 / UART/RXD	p1Do1		AIN9	模拟			OCITCK / 输入	
P1.0	p1Di0	p1Do0 / UART/TXD		AIN8	模拟	MMISO / 输入			
P0.7	p0Di7	p0Do7		AIN7	模拟	MMOSI / 输出	PWM0 / 输出		
P0.6	p0Di6	p0Do6		AIN6	模拟	MSCK / 输出	PWM1 / 输出		
P0.5	p0Di5	p0Do5		AIN5	模拟		SCSN / 输入		
P0.4	p0Di4	p0Do4		AIN4	模拟		SMISO / 输出		
P0.3	p0Di3	p0Do3		AIN3	模拟		SMOSI / 输入		
P0.2	p0Di2	p0Do2		AIN2	模拟		SSCK / 输入		
P0.1	p0Di1	p0Do1	CKLF②	AIN1	模拟				
P0.0	p0Di0	p0Do0	CKLF③ / 模拟	AIN0	模拟				

厂商推荐的 48 引脚 QFN 封装的 nRF24LE1 最小系统应用电路元器件布局和 PCB 图如图 1.2.8 和图 1.2.9 所示,电路元器件清单如表 1.2.3 所列。

注意: 元器件表中 L_2、L_3 修改为 3.9 nH,U1 修改为 nRF24LE1F16 Q48。

1.2.4 nRF24LE1 与计算机串口的连接电路

nRF24LE1 有一个串行接口(TXD 和 RXD),如 5 mm×5 mm 32 引脚 QFN 封装的 nRF24LE1 的 P0.3 和 P0.4 引脚端,该串行接口波特率可变,有两种波特率发生器可供选择。nRF24LE1 的串口与计算机串口的电路如图 1.2.10 所示。nRF24LE1 采用 3.3 V 的直流电压供电,串行接口电平转换芯片采用 MAX3232。图中,TXD_LE 和 RXD_LE 连接到 nRF24LE1 的 TXD 和 RXD。

第1章 超低功耗单片无线系统

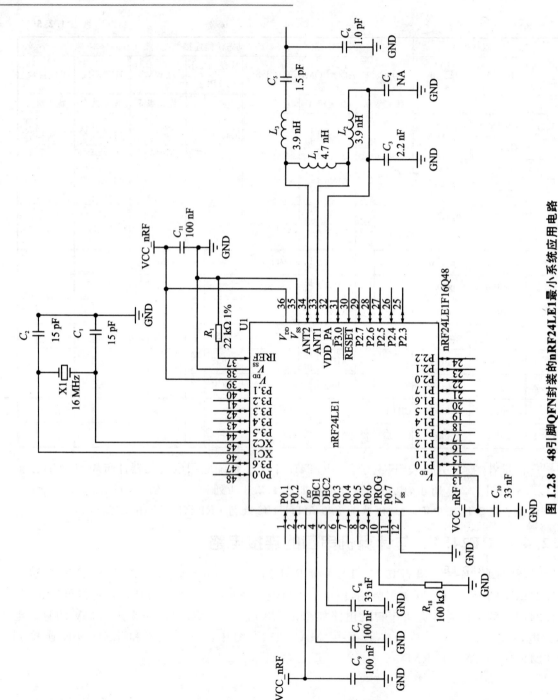

图 1.2.8 48引脚QFN封装的nRF24LE1最小系统应用电路

第1章 超低功耗单片无线系统

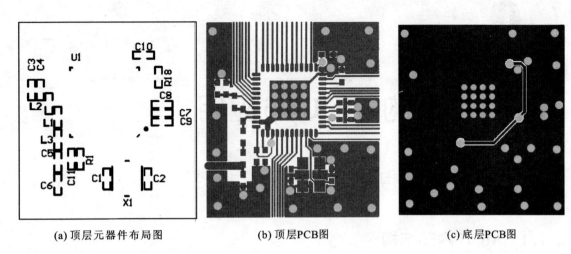

(a) 顶层元器件布局图　　(b) 顶层PCB图　　(c) 底层PCB图

图 1.2.9　48 引脚 QFN 封装的 nRF24LE1 最小系统应用电路元器件布局和 PCB 图

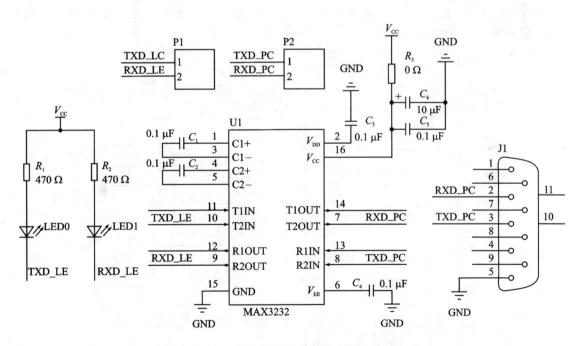

图 1.2.10　nRF24LE1 和计算机串口的连接电路

第 2 章 nRF24LE1 的 MCU 与应用

2.1 MCU 内部结构与主要特性

2.1.1 MCU 内部结构

nRF24LE1 包含一个高速 MCU（8 位单片机），MCU 内部结构方框图如图 2.1.1 所示。该 MCU 采用与标准 8051 芯片兼容的指令。MCU 指令采用单时钟指令周期，性能是标准 8051 的 8 倍（按 MIPS 性能评估）。

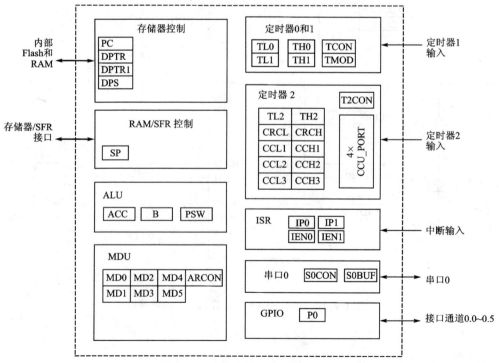

图 2.1.1 MCU 内部结构方框图

MCU 的运算逻辑单元(ALU)提供 8 位乘除、8 位带进位加、8 位带借位减,以及逻辑"与"、"或"、"异或"、"非"等逐位运算操作,所有操作都是无符号整数操作。另外,ALU 可以对寄存器递增或递减。对于累加器可以带进位或不带进位循环左移或右移,交换半字节,清 0,补码操作及十进制调整。ALU 由 ACC、B 和 PSW 3 个特殊功能寄存器来处理,操作码可以来自于累加器 ACC、寄存器 B 或外部单元,结果可以存储在累加器 ACC 或其他外部单元。PSW(程序状态字)包括进位、溢出、奇偶标志位等。

MCU 包含的片内协处理器乘除单元(MDU)可以进行 32 位除、16 位乘、移位和标准化操作。

MCU 所有指令均与标准的 8051 指令兼容。有关 MCU 的指令及功能的更多内容请登录 www.nordicsemi.com,查询 nRF24LE1 Ultra-low Power Wireless System On-Chip Solution Preliminary Product Specification v1.6,并参考有关的 8051 指令资料。

2.1.2 MCU 主要特性

下面介绍 nRF24LE1 的 MCU 主要特性。

① 控制单元:
- 8 位指令解码;
- 高速指令周期(12 倍于工业标准 80C51)。

② 运算逻辑单元:
- 8 位算术和逻辑操作;
- 布尔操作;
- 8×8 位乘法和 8/8 位除法。

③ 乘除法单元:
- 16×16 位乘法;
- 32/16 位和 16/16 位除法;
- 32 位标准操作;
- 32 位左移/右移。

④ 3 个 16 位定时器/计数器:
- 与 80C51 相似的定时器 0 和 1;
- 与 80515 相似的定时器 2。

⑤ 比较器/捕获器单元,定时器 2 专用:
- 软件控制捕获;
- 4 个 16 位比较寄存器用来作为脉宽调制;
- 4 个外部捕获输入用来作为脉冲宽度测量;
- 16 位重载寄存器用来作为脉冲发生器。

⑥ 全双工串口:
- 串口0(与80C51类似);
- 同步模式,固定波特率;
- 8位UART模式,可变波特率;
- 9位UART模式,固定波特率;
- 9位UART模式,可变波特率;
- 波特率发生器。

⑦ 中断控制:
- 13个中断源,4个优先级。

⑧ 存储器接口:
- 16位地址总线;
- 双数据指针,用于快速数据块传输。

⑨ 硬件支持软件调试。

2.2 存储器和I/O结构

2.2.1 存储器映射

nRF24LE1的MCU片内协处理器乘除单元(MDU)有64 KB的程序和数据存储器空间、256字节内部数据存储器(IRAM)和128字节特殊功能寄存器(SFR)。

nRF24LE1的存储器有一个默认的设置:16 KB可编程存储器(Flash)、1 KB数据存储器(SRAM)及2块(1 KB标准寿命和512字节长寿命)非易失性存储器(Flash)。默认的存储器映射图如图2.2.1所示。编程和NVM(非易失性存储器)存储块的大小可以根据应用需要重新配置。

内部数据存储器(IRAM)的低128字节包含工作寄存器(0x00~0x1F)和可位寻址存储器(0x20~0x2F)。上面一半只能间接寻址。IRAM的低32字节分为4组,每组包括8个寄存器(R0~R7)。由PSW寄存器中的两位来确定使用哪一组。下面的16字节存储器组成一个位寻址存储器,通过位地址0x00~0x7F访问。

2.2.2 PDATA存储器寻址

nRF24LE1支持页数据存储(PDATA)到数据地址空间寻址。每一页(256字节)可以通过寄存器R0和R1(@R0,@R1)来间接寻址。由MPAGE寄存器来控制PDATA页的首地址,如表2.2.1所列。

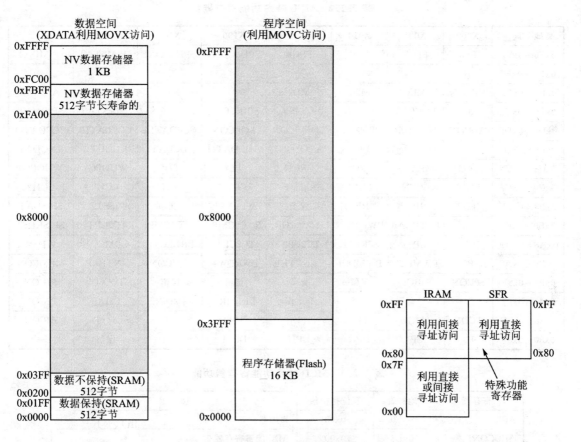

图 2.2.1　默认的存储器映射图

表 2.2.1　MPAGE 寄存器

地　址	位	R/W	功　能	复位值
0xC9	[7:0]	R/W	PDATA 页起始地址	0x00

MPAGE 设置 16 位地址空间的高 8 位。例如,设置 MPAGE 为 0x80,则 PDATA 的起始地址是 0x8000。

2.2.3　MCU 特殊功能寄存器

MCU 特殊功能寄存器及其功能分别如表 2.2.2 和表 2.2.3 所列。表 2.2.2 中,未定义的部分禁止读/写,X000 列的 B 寄存器可以字节和位寻址,其他只能字节寻址。

第 2 章 nRF24LE1 的 MCU 与应用

表 2.2.2 MCU 特殊功能寄存器

地址	X000	X001	X010	X011	X100	X101	X110	X111
0xF8~0xFF	FSR	FPCR	FCR	FDCR	SPIMCON0	SPIMCON1	SPIMSTAT	SPIMDAT
0xF0~0xF7	B	—	—	—	—	—	—	—
0xE8~0xEF	RFCON	MD0	MD1	MD2	MD3	MD4	MD5	ARCON
0xE0~0xE7	ACC	W2CON1	W2CON0	保留	SPIRCON0	SPIRCON1	SPIRSTAT	SPIRDAT
0xD8~0xDF	ADCON	W2SADR	W2DAT	COMPCON	POFCON	CCPDATIA	CCPDATIB	CCPDATO
0xD0~0xD7	PSW	ADCCON3	ADCCON2	ADCCON1	ADCDATH	ADCDATL	RNGCTL	RNGDAT
0xC8~0xCF	T2CON	MPAGE	CRCL	CRCH	TL2	TH2	WUOPC1	WUOPC0
0xC0~0xC7	IRCON	CCEN	CCL1	CCH1	CCL2	CCH2	CCL3	CCH3
0xB8~0xBF	IEN1	IP1	S0RELH	保留	SPISCON0	SPISCON1	SPISSTAT	SPISDAT
0xB0~0xB7	P3	RSTREAS	PWMCON	RTC2CON	RTC2CMP0	RTC2CMP1	RTC2CPT00	SPISRDSZ
0xA8~0xAF	IEN0	IP0	S0RELL	RTC2CPT01	RTC2CPT10	CLKLFCTRL	OPMCON	WDSV
0xA0~0xA7	P2	PWMDC0	PWMDC1	CLKCTRL	PWRDWN	WUCON	INTEXP	MEMCON
0x98~0x9F	S0CON	S0BUF	保留	保留	保留	保留	P0CON	P1CON
0x90~0x97	P1	任意	DPS	P0DIR	P1DIR	P2DIR	P3DIR	P2CON
0x88~0x8F	TCON	TMOD	TL0	TL1	TH0	TH1	保留	P3CON
0x80~0x87	P0	SP	DPL	DPH	DPL1	DPH1	保留	—

表 2.2.3 MCU 特殊功能寄存器功能

寄存器	地址	复位初始值	功能
ACC	0xE0	0x00	累加器
ADCCON1	0xD3	0x00	ADC 配置寄存器 1
ADCCON2	0xD2	0x00	ADC 配置寄存器 2
ADCCON3	0xD1	0x00	ADC 配置寄存器 3
ADCDATH	0xD4	0x00	ADC 数据高字节
ADCDATL	0xD5	0x00	ADC 数据低字节
ARCON	0xEF	0x00	算术控制寄存器
B	0xF0	0x00	B 寄存器
CCEN	0xC1	0x00	比较/捕获使能寄存器
CCH1	0xC3	0x00	比较/捕获寄存器 1,高字节
CCH2	0xC5	0x00	比较/捕获寄存器 2,高字节
CCH3	0xC7	0x00	比较/捕获寄存器 3,高字节
CCL1	0xC2	0x00	比较/捕获寄存器 1,低字节
CCL2	0xC4	0x00	比较/捕获寄存器 2,低字节
CCL3	0xC6	0x00	比较/捕获寄存器 3,低字节

续表 2.2.3

寄存器	地　址	复位初始值	功　能
CCPDATIA	0xDD	0x00	加密/解密协处理器数据输入寄存器 A
CCPDATIB	0xDE	0x00	加密/解密协处理器数据输入寄存器 B
CCPDATO	0xDF	0x00	加密/解密协处理器数据输出寄存器
CLKLFCTRL	0xAD	0x07	32 kHz (CLKLF)控制寄存器
CLKCTRL	0xA3	0x00	时钟控制寄存器
COMPCON	0xDB	0x00	比较器控制寄存器
CRCH	0xCB	0x00	比较/重载/捕获寄存器,高字节
CRCL	0xCA	0x00	比较/重载/捕获寄存器,低字节
DPH	0x83	0x00	数据指针 0 高字节
DPL	0x82	0x00	数据指针 0 低字节
DPH1	0x85	0x00	数据指针 1 高字节
DPL1	0x84	0x00	数据指针 1 低字节
DPS	0x92	0x00	数据指针选择寄存器
FCR	0xFA	—	Flash 命令寄存器
FDCR	0xFB	—	Flash 数据配置寄存器
FPCR	0xF9	—	Flash 保护配置寄存器
FSR	0xF8	—	Flash 状态寄存器
IEN0	0xA8	0x00	中断使能寄存器 0
IEN1	0xB8	0x00	中断优先级寄存器/使能寄存器 1
INTEXP	0xA6	0x01	中断扩展寄存器
IP0	0xA9	0x00	中断优先级寄存器 0
IP1	0xB9	0x00	中断优先级寄存器 1
IRCON	0xC0	0x00	中断请求控制寄存器
MD0	0xE9	0x00	乘法/除法寄存器 0
MD1	0xEA	0x00	乘法/除法寄存器 1
MD2	0xEB	0x00	乘法/除法寄存器 2
MD3	0xEC	0x00	乘法/除法寄存器 3
MD4	0xED	0x00	乘法/除法寄存器 4
MD5	0xEE	0x00	乘法/除法寄存器 5
MEMCON	0xA7	0x00	存储器配置寄存器
MPAGE	0xC9	0x00	PDATA 页的起始地址
OPMCON	0xAE	0x00	工作模式控制寄存器
P0	0x80	0xFF	Port 0 的值
P0CON	0x9E	0x10	Port 0 配置寄存器
P0DIR	0x93	0xFF	Port 0 引脚方向控制寄存器

第 2 章 nRF24LE1 的 MCU 与应用

续表 2.2.3

寄存器	地 址	复位初始值	功 能
P1	0x90	0xFF	Port 1 的值
P1CON	0x9F	0x10	Port 1 配置寄存器
P1DIR	0x94	0xFF	Port 1 引脚方向控制寄存器
P2	0xA0	0xFF	Port 2 的值
P2CON	0x97	0x10	Port 2 配置寄存器
P2DIR	0x95	0xFF	Port 2 引脚方向寄存器
P3	0xB0	0xFF	Port 3 的值
P3CON	0x8F	0x10	Port 3 配置寄存器
P3DIR	0x96	0xFF	Port 3 引脚方向控制寄存器
POFCON	0xDC	0x00	电源故障比较器配置寄存器
PSW	0xD0	0x00	程序状态字
PWMCON	0xB2	0x00	PWM 配置寄存器
PWMDC0	0xA1	0x00	PWM 通道 0 占空系数
PWMDC1	0xA2	0x00	PWM 通道 1 占空系数
PWRDWN	0xA4	0x00	低功耗控制
RFCON	0xE8	0x02	RF 射频收发控制寄存器
RNGCTL	0xD6	0x40	随机数发生器控制寄存器
RNGDAT	0xD7	0x00	随机数发生器数据寄存器
RSTREAS	0xB1	0x00	复位原因寄存器
RTC2CMP0	0xB4	0xFF	RTC2 比较值寄存器 0
RTC2CMP1	0xB5	0xFF	RTC2 比较值寄存器 1
RTC2CON	0xB3	0x00	RTC2 配置寄存器
RTC2CPT00	0xB6	0x00	RTC2 捕获值寄存器 00
RTC2CPT01	0xAB	0x00	RTC2 捕获值寄存器 01
RTC2CPT10	0xAC	0x00	RTC2 捕获值寄存器 10
S0BUF	0x99	0x00	串口 0,数据缓冲
S0CON	0x98	0x00	串口 0,控制寄存器
S0RELH	0xBA	0x03	串口 0,重载寄存器,高字节
S0RELL	0xAA	0xD9	串口 0,重载寄存器,低字节
SP	0x81	0x07	栈指针
SPIMCON0	0xFC	0x02	SPI 主机配置寄存器 0
SPIMCON1	0xFD	0x0F	SPI 主机配置寄存器 1
SPIMDAT	0xFF	0x00	SPI 主机数据寄存器
SPIMSTAT	0xFE	0x03	SPI 主机状态寄存器
SPIRCON0	0xE4	0x01	RF 射频收发 SPI 主机配置寄存器 0

续表 2.2.3

寄存器	地址	复位初始值	功能
SPIRCON1	0xE5	0x0F	RF 射频收发 SPI 主机配置寄存器 1
SPIRDAT	0xE7	0x00	RF 射频收发 SPI 主机数据寄存器
SPIRSTAT	0xE6	0x03	RF 射频收发 SPI 主机状态寄存器
SPISCON0	0xBC	0xF0	SPI 从机配置寄存器 0
SPISCON1	0xBD	0x0F	SPI 从机配置寄存器 1
SPISDAT	0xBF	0x00	SPI 从机数据寄存器
SPISRDSZ	0xB7	0x3F	SPI 从机接收数据长度寄存器
SPISSTAT	0xBE	0x03	SPI 从机状态寄存器
T2CON	0xC8	0x00	定时器 2 控制寄存器
TCON	0x88	0x00	定时器/计数器控制寄存器
TH0	0x8C	0x00	定时器 0,高字节
TH1	0x8D	0x00	定时器 1,高字节
TH2	0xCD	0x00	定时器 2,高字节
TL0	0x8A	0x00	定时器 0,低字节
TL1	0x8B	0x00	定时器 1,低字节
TL2	0xCC	0x00	定时器 2,低字节
TMOD	0x89	0x00	定时器模式寄存器
W2CON0	0xE2	0x80	2 线配置寄存器 0
W2CON1	0xE1	0x00	2 线配置寄存器 1/状态寄存器
W2DAT	0xDA	0x00	2 线数据寄存器
W2SADR	0xD9	0x00	2 线从地址寄存器
ADCON	0xD8	0x00	串口 0 波特率选择寄存器(仅使用 ADCON[7]位)
WDSW	0xAF	0x00	看门狗初始值寄存器
WUCON	0xA5	0x00	看门狗配置寄存器
WUOPC0	0xCF	0x00	引脚端唤醒配置 0
WUOPC1	0xCE	0x00	引脚端唤醒配置 1

在 MCU 特殊功能寄存器中,包括:

① 累加器 ACC。MCU 的大多数指令均使用累加器来保存操作数和存储操作结果。累加器的助记符是 A,而不是 ACC。

② 寄存器 B。在乘除指令时使用,可用来保存临时数据。

③ 程序状态寄存器 PSW。所包含的状态位反映当前 MCU 的状态。

④ 堆栈指针 SP。它是一个专用寄存器,指出堆栈顶部在内部 RAM 空间中的位置,用来存储程序执行中断或子程序返回的地址。在执行 PUSH 或 CALL 指令时,SP 加 1;在执行 POP 或 RET(I)指令时,SP 减 1(总是指向栈的顶端)。

⑤ 数据指针 DPH、DPL。数据指针寄存器可以通过 DPL 和 DPH 来访问。当前的数据指针由 DPS 寄存器选择。这些寄存器用来保存在间接寻址模式 16 位地址,如 MOVX、MOVC 或 JMP 指令。它们可构成一个 16 位的寄存器或 2 个独立的 8 位寄存器。DPH 用来保存间接寻址模式 16 位地址的高 8 位,DPL 用来保存间接寻址模式 16 位地址的低 8 位。通常用来访问外部程序或数据存储器,例如"MOVC A,@A+DPTR;"或"MOV A,@DPTR;"。

⑥ 数据指针 1 DPH1、DPL1。地址为 0x84、0x85。数据指针寄存器 1 可以通过 DPL1 和 DPH1 来访问。当前的数据指针由 DPS 寄存器选择。它们可以构成一个 16 位的寄存器或者 2 个独立的 8 位寄存器。DPH1 用来间接寻址模式 16 位地址的高 8 位,DPL1 用来间接寻址模式 16 位地址的低 8 位。通常用来访问外部程序或数据存储器,例如"MOVC A,@A+DPTR;"或"MOV A,@DPTR;"。数据指针 1 是标准 8051 结构的扩展,用来加速数据块的传输。

⑦ 数据指针选择 DPS。MCU 包含有两组数据指针,都可以用来作为间接寻址的 16 位源地址。DPS 寄存器用于选择哪一组数据指针有效。

⑧ 寄存器 PCON 用来控制程序存储器写模式和串口 0 波特率加倍。

有关 MCU 特殊功能寄存器各位的定义与功能的更多内容请登录 www.nordicsemi.com,查询 nRF24LE1 Ultra-low Power Wireless System On-Chip Solution Preliminary Product Specification v1.6。

2.3 Flash 存储器

2.3.1 Flash 存储器特性

nRF24LE1 中的 Flash 块分为 16 KB 的通用程序空间和 1.5 KB 非易失数据存储器空间,如图 2.3.1 所示。MCU 可读/写 Flash 存储器,在特殊情况下(如执行固件升级时)MCU 还可进行擦/写操作。Flash 存储器也可通过外部的从 SPI 接口来配置和编程。编程后,利用程序保护功能可禁止外部接口的读/写操作。

nRF24LE1 的 Flash 存储器具有如下特性:
- 16 KB 程序存储器;
- 1 KB NV 非易失数据存储器;
- 程序存储器和非易失数据存储器的页面大小均为 512 字节;
- 2 页(各 256 字节)的长擦/写寿命存储器;
- 32 页主 Flash 块加 1 个信息页(InfoPage);
- 普通 Flash 块可最少擦/写 1 000 次;
- 扩展 Flash 块可最少擦/写 20 000 次;

- 可直接对 SPI 进行编程；
- 可配置 MCU 写保护；
- 可配置读出保护；
- 硬件设计支持软件升级。

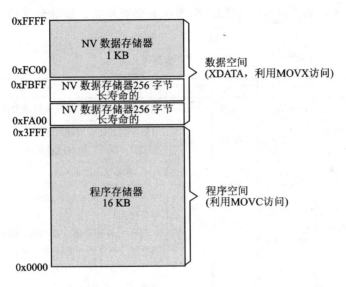

图 2.3.1　nRF24LE1 中的 Flash 块结构

2.3.2　Flash 存储器配置

Flash 存储器块提供了 MCU 可存储程序代码和应用数据的非易失存储空间（NVM），2 页 256 字节长寿命非易失数据存储空间可擦/写 20 000 次。Flash 存储器可由 MCU 通过程序和数据空间来操作。

在正常模式下，配置和设置存储器特性由存储在独立的信息页中的数据来定义。在芯片上电复位时，信息页中的配置被读出，并存储在存储器配置特殊寄存器中。

片内 Flash 存储器分为 2 块：16 KB+1.5 KB 非易失存储器的主块（MB）和 512 字节的信息页（IP）。存储器配置信息存储在信息页中，可以配置下面的信息：

- 将程序空间划分为保护块和非保护块 2 块（针对 MCU 的擦/写操作）；
- 禁止 Flash 外部接口 SPI 和 HW 调试的读/写访问；
- 使能硬件调试功能。

所有 Flash 存储器的配置必须通过外部 SPI 接口来完成。编程时，配置信息将存储在信息页中，并在每次复位/启动时被读出到相应的特殊功能寄存器。

第 2 章　nRF24LE1 的 MCU 与应用

1. 信息页

信息页在 Flash 存储器中是一个独立的 512 字节页,包含了 Nordic 系统的调整参数和 Flash 存储器的配置选项。任何对 Flash 存储器配置的改变必须更新此页的信息。信息页内容如表 2.3.1 所列。

注意：信息页区用来存储 nRF24LE1 的系统和调整参数。擦除该区的内容将会导致器件性能和特性的改变。

表 2.3.1　信息页内容

信息页数据	符号	大小	地址	注释
器件系统参数	DSYS	32 字节	0x00	保留给器件使用。不允许擦除与修改
无保护的页数；NUPP(被保护区的起始页地址)	NUPP	1 字节	0x20	在启动时读出寄存器 FPCR。 NUPP=0xFF：所有的页无保护
保留		2 字节	0x21	保留,必须为 0xFF
Flash 主块读保护	RDISMB	1 字节	0x23	禁止外部接口对 Flash 主块的访问。 字节值： 0xFF—允许外部接口对 Flash 主块的访问； 其他值—不允许外部接口对 Flash 主块的读/擦除/写。仅能够被 SPI 指令 RDISMB 改变一次。仅能够被 SPI 指令 ERASE ALL 复位
使能硬件调试	ENDEBUG	1 字节	0x24	使能硬件调试和 JTAG 接口。 字节值： 0xFF—不使能硬件调试； 其他值—使能硬件调试和 JTAG 接口
保留	—	486 字节	0x25	保留,必须是 0xFF

(1) DSYS(器件系统参数)

这个信息页区被 nRF24LE1 用来存储核心数据,如调整参数。擦除或改变该区的内容将可能严重影响系统的性能。可以影响此区操作的 SPI 指令有 ERASE ALL、ERASE PAGE 和 PROGRAM(当 FSR 寄存器中的 INFEN 位 1 时,程序对此区 Flash 地址的操作)。如果要使用 ERASE ALL 的 SPI 指令,必须先读出信息页中的内容并存储,然后在 ERASE ALL 指令执行完后,再写回到 nRF24LE1 的信息页中。

(2) NUPP(无保护的页数)

Flash 区可以划分为非保护区和保护区。MCU 对保护区只能进行读操作,但是可以通过外部 SPI 接口进行读/写及擦除操作。保护功能是防止 MCU 对代码空间(程序存储器空间)进行非法擦/写操作,这个功能主要用于固件升级。

Flash 主块的代码空间分为 32 页,每页 512 字节。不改变 NUPP 字节(NUPP=0xFF)即 32 页代码空间为非保护状态,MCU 可以对任何部分进行擦/写。如果 NUPP<32,Flash 主块的代码空间将被划分为非保护区和保护区 2 块,其中非保护区页数=NUPP,保护区页数为 31-NUPP。例如,NUPP=12,将有 12 个非保护页(0~11 页)和 20 个保护页(12~31 页)。

如果已经将 Flash 主块划分为 2 块,那么 FSR 寄存器中 STP 位的值将决定 MCU 从哪里开始执行程序代码。通常情况下,STP 设为 0,而程序代码将从 0x0000 开始执行。如果 STP 为 1,则程序将从保护区开始执行代码。STP 位在复位/器件启动时设置,如果此时在 Flash 数据存储器的最高 16 个地址内 1 的个数为奇数,STP 将被设为 1,如图 2.3.2 所示。

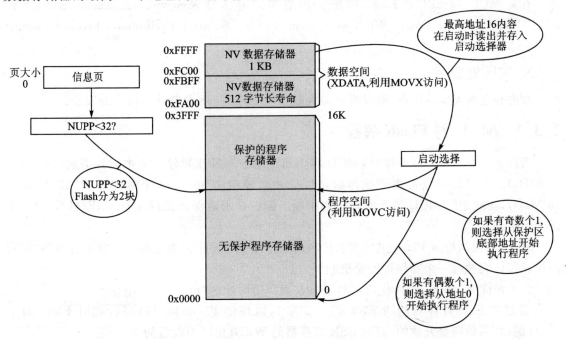

图 2.3.2 Flash 主块保护区

采用触发使能代码从保护区开始执行看起来有些繁琐,但是这样的操作确保了固件升级时的安全性。

(3) RDISMB(Flash 主块读保护)

将此字节的内容由 0xFF 改为其他任何值;将禁止 SPI 和其他外部接口对 Flash 主块进行任何访问,而只允许对信息页进行读操作。此字节的内容可通过 RDISMB SPI 指令进行改变,由于此指令将切断 SPI 访问 Flash 主块,因此必须在 Flash 编程后才可以发送此指令到 nRF24LE1。可使 SPI 重新访问 Flash 主块的指令是 ERASE ALL。

注意:ERASE ALL 将擦除整个信息页。使用 ERASE ALL 指令前,如果没有读出/存储信息页数据,并在执行指令后将 DSYS 写回信息页,将会导致器件失去功能!

(4) ENDEBUG(使能硬件调试)

将此字节的内容由 0xFF 改为其他任何值,将使能片内硬件调试功能和 JTAG 调试接口。片内硬件调试功能将改变器件的引脚分配定义,并需要 nRFprobe(Nordic 提供的调试工具)或 FS2(第三方提供的调试工具)硬件调试工具来配合使用。

2. 存储器配置特殊功能寄存器

引导程序将 Flash 信息页的内容导入存储器配置特殊功能寄存器中。同样,配置的特殊功能寄存器内容将用来与 SPI 和 MCU 进行接口。

有关 MCU 和 SPI 对 Flash 的配置的更多内容请登录 www.nordicsemi.com,查询 nRF24LE1 Ultra-low Power Wireless System On-Chip Solution Preliminary Product Specification v1.6。

3. 欠压处理

片内有电源欠压检测器,确保当电源失效(POF)时 Flash 存储器写操作的安全。

2.3.3 MCU 对 Flash 编程

当启动一个 Flash 写操作时,MCU 将为每个字节写操作暂停 740 个时钟周期(46 μs @ 16 MHz)。当启动一个页擦除操作时,MCU 将暂停最多 360 000 个时钟周期(22.5 ms @ 16 MHz),在此期间 MCU 将不响应任何中断。固件必须确保页擦除不被 nRF24LE1 的正常操作干扰。

MCU 可以执行对 Flash 主块非保护区的页擦除和写操作。为了预防非预期有损的擦/写操作,MCU 将执行一个写保护安全机制。

为了允许 Flash 擦/写操作,MCU 必须按照以下时序运行:

① 设置 FSR 寄存器的 WEN(位[5])为 1,以使能 Flash 擦/写访问,此时 Flash 对于 MCU 的擦/写访问是开放的,直到 FSR 寄存器的 WEN(位[5])设置为 0。

② 要更新 Flash 存储器的内容,首先必须擦除,擦除操作只能对整页进行。为了擦除一页,则必须写页地址(0~31)到 FCR 寄存器。

③ 设置 PCON 寄存器的 PMW(位[4])为 1,以使能程序存储器写模式。

④ MCU 通过正常的存储器写操作即可对 Flash 进行编程。字节是按指定地址单个写入的(没有自动递增功能)。

当编程代码是从 Flash 执行时,擦/写操作的时间是自身控制的,同时 CPU 将停止工作直到操作结束。当编程代码是从 XDATA RAM 执行时,代码必须等待所有操作结束。这可以通过查询 FSR 寄存器的 RDYN 位是否变低(0)或等待循环。在擦/写操作完成前不要将 WEN 置 0。存储器地址等于 Flash 地址。

```c
typedef unsigned char uint8_t;
typedef unsigned int  uint16_t;
/****************************************************************
/函数名称：flash_page_erase()
/函数功能：擦除指定的 Flash 页
/输入参数：pn 为要擦除的 Flash 页
/返回参数：无
****************************************************************/
void flash_page_erase(uint8_t pn)
{
    F0 = EA;
    EA = 0;
    WEN = 1;
    FCR = pn;
    while(RDYN == 1);
    WEN = 0;
    EA = F0
}
/****************************************************************
/函数名称：flash_byte_write()
/函数功能：往指定的地址单元 a 写入数据 b
/输入参数：a 为 Flash 地址，b 为数据
/返回参数：无
****************************************************************/
void flash_byte_write(uint16_t a,uint8_t b)
{
    uint8_t xdata * data pb;
    F0 = EA;
    EA = 0;
    WEN = 1;
    pb = (uint8_t xdata *)a;
    *pb = b;
    while(RDYN == 1);
    WEN = 0;
    EA = F0;
}
/****************************************************************
/函数名称：flash_bytes_write()
/函数功能：以 a 为起始地址连续写入 p 指向的 n 个数据
```

第 2 章　nRF24LE1 的 MCU 与应用

/输入参数: a 为 Flash 地址,p 为数据指针,n 为写入的数据长度
/返回参数: 无
**/
```
void flash_bytes_write(uint16_t a,uint8_t *p,uint16_t n)
{
    uint8_t xdata * data pb;
    F0 = EA;
    EA = 0;
    WEN = 1;
    pb = (uint8_t xdata *)a;
    while(n--)
    {
        *pb++ = *p++;
        while(RDYN == 1);
    }
    WEN = 0;
    EA = F0;
}
```
/***
/函数名称: flash_byte_read()
/函数功能: 从地址单元 a 中读取一字节
/输入参数: a 为 Flash 地址
/返回参数: 返回地址单元 a 中的数据
**/
```
uchar flash_byte_read(uint16_t a)
{
    uint8_t xdata *pb = (uint8_t xdata *)a;
    return *pb;
}
```
/***
/函数名称: flash_bytes_read()
/函数功能: 从地址单元 a 连续读取 n 个数据存到 P 指向的存储单元
/输入参数: a 为 Flash 地址,p 为数据指针,n 为将读的数据个数
/返回参数: 无
**/
```
void flash_bytes_read(uint16_t a,uint8_t *p,uint16_t n)
{
    uint8_t xdata *pb = (uint8_t xdata *)a;
    while(n--)
```

```
    {
        *p = *pb;
        pb++;
        p++;
    }
}
```

2.3.4 通过 SPI 对 Flash 编程

片内的 Flash 可以通过 SPI 接口来编程,该编程接口使用 8 位指令寄存器及一组编程和配置 Flash 存储器的指令/命令。

1. SPI 从接口

为编程存储器,须使用 SPI 从接口。当引脚 PROG=1,且复位引脚保持为非有效时,SPI 从接口连接到 Flash 存储器。当 PROG 引脚端置高时,被选择的 nRF24LE1 通用 I/O 引脚端自动配置为从模式,如表 2.3.2 所列。

注意:在激活 PROG 引脚后,在输入第一个 Flash 命令前,必须等待至少 1.5 ms。

表 2.3.2 不同封装的 nRF24LE1 的 Flash 从接口定义

符号	24 引脚 4 mm×4 mm	32 引脚 5 mm×5 mm	48 引脚 7 mm×7 mm
FCSN	P0.5	P1.1	P2.0
FMISO	P0.4	P1.0	P1.6
FMOSI	P0.3	P0.7	P1.5
FSCK	P0.2	P0.5	P1.2

2. Flash 的操作指令

编程接口使用 8 位指令寄存器和一组指令/命令去编程和配置 Flash 存储器,如表 2.3.3 所列。

表 2.3.3 Flash 的操作指令

指令	指令格式	地址	#数据字节	指令操作
WREN	0x06	NA	0	设置 Flash 写使能锁存。 FSR 寄存器的 WEN 位
WRDIS	0x04	NA	0	复位 Flash 写使能锁存。 FSR 寄存器的 WEN 位
RDSR	0x05	NA	1	读 Flash 状态寄存器(FSR)
WRSR	0x01	NA	1	写 Flash 状态寄存器(FSR)。 注意:FSR 中的 DBG 位仅能够被 MCU 置位

第 2 章　nRF24LE1 的 MCU 与应用

续表 2.3.3

指　令	指令格式	地　址	#数据字节	指令操作
READ	0x03	2 字节。读操作的第 1 个 Flash 地址	1～18432xx	从 Flash 存储器读数据
PROGRAM	0x02	2 字节。写操作的第 1 个 Flash 地址	1～18432xx	写数据到 Flash 存储器。 注意：WEN 必须置位
ERASE PAGE	0x52	2 字节。被删除的第 1 个页地址	0	擦除指定地址页。 注意：WEN 必须置位
ERASE ALL	0x62①	NA	0	擦除在 Flash 主块中的所有页和信息页。 注意：WEN 必须置位
RDFPCR	0x89	NA	1	读 Flash 保护配置寄存器 FPCR
RDI SMB	0x85	NA	0	使能 Flash 读出保护。 注意：WEN 必须置位
ENDEBUG	0x86	NA	0	使能硬件调试功能。 注意：WEN 必须置位。只能够操作一次

① 信息页区用来存储 nRF24LE1 的系统和调整参数。擦除这个区的内容将会改变器件的性能和特性。在执行 ERASE ALL 指令前，信息页的 DSYS 区的数据必须读取和存储，而在完成擦除操作后，必须将已存储的 DSYS 区数据写回到 Flash 的信息页。

3. SPI 接口信号

SPI 接口信号的传输示意图如图 2.3.3 和图 2.3.4 所示。图中：Cx 为 SPI 指令位；Ax 为 Flash 地址，低字节到高字节，字节的高位先；Dx 为 SPI 数据位，低字节到高字节，字节的高位先。

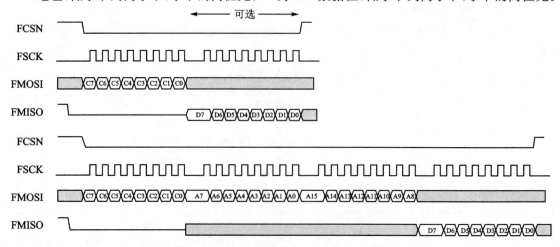

图 2.3.3　SPI 读操作时序（直接和地址指令）

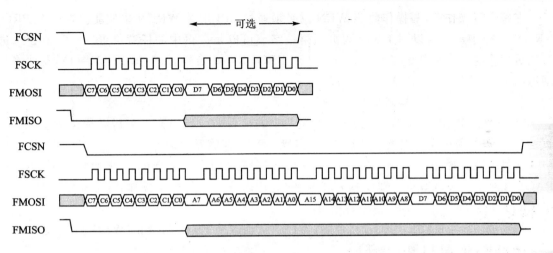

图 2.3.4　SPI 写操作时序(直接和地址指令)

(1) WREN/WRDIS(Flash 写使能/禁止)

WREN 和 WRDIS 都是一个不包含数据的单字节 SPI 指令。

SPI 指令 WREN 和 WRDIS 设置和清除 FSR 寄存器中的 WEN 位,以使能/禁止整个 Flash 块写操作。

器件上电后为写禁止状态,并在每个擦/写 SPI 指令后(FCSN 置高)自动回到写禁止状态。因此,每个通过 SPI 接口的擦/写指令必须通过一个 WREN 指令呈现。

(2) RDSR/WRSR(读/写 Flash 状态寄存器)

SPI 指令 RDSR 和 WRSR 可读/写 Flash 状态寄存器 FSR,指令均为 1 字节并跟随一个包含有 FSR 内容的数据字节。

(3) READ(读)

SPI 指令 Read 读出一个指定地址的 Flash 主块内容,必须跟随指示读出操作的 2 字节地址,见图 2.3.3。如果 FSR 寄存器的 INFEN 位使能,则读操作将会读出信息页的内容。

如果 FCSN 线在第一个数据字节读出后保持有效,则读指令可以扩展,即地址自动递增而数据连续读出。当到达最高地址时,内部地址计数器将会开始循环,允许在一个连续的读指令中读出整个存储器。

读出 Flash 主块内容的操作仅在 FSR 寄存器的读禁止位(即 RDISMB 位)未被置位时允许。

(4) PROGRAM(编程)

SPI 指令 PROGRAM 编程指定地址的 Flash 主块中的内容,指令必须跟随指示写操作的 2 字节起始地址,见图 2.3.4。如果 FSR 寄存器的 INFEN 位被使能,则写操作将从信息页开始。

在每个写操作前,写操作使能 WEN 位必须通过 SPI 指令 WREN 来使能。在一个 PROGRAM 指令中最多可以写 1 KB(两页),第一字节可以是一页中的任意位置。一字节在未进行整页擦除前不能重新编程。当执行完一个 PROGRAM 指令后(引脚 FCSN=1),器件将自动返回 Flash 写禁止状态(WEN=0)。

(5) ERASE PAGE(擦除页)

SPI 指令 ERASE PAGE 擦除 Flash 主块中的一页(512 字节),指令后必须跟随一字节的页地址(0~31 为程序存储器页地址,32~35 为 NVM 存储器页地址)。

在每个擦除操作前,写操作使能 WEN 位必须通过 SPI 指令 WREN 来使能。在 ERASE PAGE 指令后,当 FCSN 引脚设置为高电平时,片内开始执行擦除操作;在擦除操作期间,除了 RDSR 指令外的其他 SPI 指令将被忽略。

当完成一个 ERASE PAGE 指令后,器件将自动返回 Flash 写禁止(WEN=0)状态。

(6) ERASE ALL(擦除全部)

SPI 指令 ERASE ALL 擦除 Flash 主块的所有页(程序代码页和 NVM 页)和信息页,该指令是 1 字节不带数据的指令。

在擦除操作前,写操作使能 WEN 位必须通过 SPI 指令 WREN 来使能。在 ERASE ALL 指令后,当 FCSN 引脚设置为高电平时,片内开始执行擦除操作;在擦除操作期间,除了 RDSR 指令外的其他 SPI 指令将被忽略。

如果 FSR 寄存器的 INFEN 位在执行 ERASE ALL 指令前置高,则信息页和 Flash 主块都将被擦除,否则将只有 Flash 主块被擦除。

当完成一个 ERASE ALL 指令后,器件将自动返回 Flash 写禁止(WEN=0)状态。

(7) RDFPCR(读 Flash 保护配置寄存器)

SPI 指令 RDFPCR 读出 Flash 保护配置寄存器(FPCR)内容,寄存器内容包含 Flash 主块写保护页的配置。指令后面跟随 1 字节数据。

(8) RDISMB(使能主块读禁止)

SPI 指令 RDISMB 使能 Flash 读保护。指令禁止所有通过外部接口(SPI 及 JTAG 调试口)对 Flash 主块的读/擦/写访问,也禁止对信息页的擦/写操作,但允许读出信息页内容。这将保护片内程序代码和数据的内容不被外部接口获取。

在 RDISMB 指令前,写操作使能 WEN 位必须通过 SPI 指令 WREN 来使能。一旦 RDISMB 指令发送后,SPI 将失去所有对 Flash 的连接和控制。因此,必须确保此指令是对 Flash 编程操作的最后一组指令。该指令是 1 字节不带数据的指令。

(9) ENDEBUG(使能硬件调试)

SPI 指令 ENDEBUG 使能片上硬件调试,同时使能 JTAG 接口。

在操作前,写操作使能 WEN 位必须通过 SPI 指令 WREN 来使能,硬件调试功能使能后只能通过 ERASE ALL 指令来禁止。该指令是 1 字节不带数据的指令。

2.3.5 硬件支持固件升级

当一部分 Flash 存储器配置为 MCU 写保护(FPCR、NUPP)时，nRF24LE1 将从保护区开始启动，存储器映射实际上发生改变，以确保固件升级的安全。编程后程序代码空间的非保护区和保护区示例如图 2.3.5 所示。

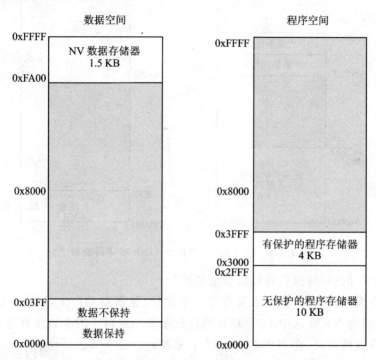

图 2.3.5　带 4 KB 保护区的 Flash 存储器映射示例

重启动后，地址映射将发生改变，保护区从 0x0000 开始往上映射，如图 2.3.6 所示。

非保护区在数据空间是可用的，这使得升级更为容易。

注意：SRAM 块的地址映射从 0x8000 开始不受 MEMCON 位[2]的影响，这个功能可用于空中固件升级。

固件升级的方法如下：

- 应用程序运行在非保护区，固件升级的驻留程序放在保护区；
- 通过通信器件启动一个空中固件升级；
- MCU 设置 WEN；
- 写一个 0 到 NV 数据存储器中 16 个最高地址位字节中的第一位，使该区逻辑 1 的个数为奇数；
- 此时系统可以复位，由于有 STP 位，因此将从保护区开始重新启动；

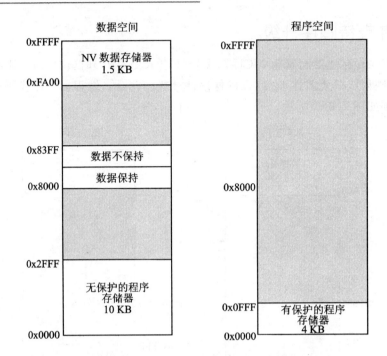

图 2.3.6　带 4 KB 保护区的 Flash 存储器映射

- 在非保护区的擦/写操作将可以安全地执行；
- 万一电源失效，或在升级完成前有另一个复位/重新启动产生，MCU 仍将从保护区开始执行，因为 NVM 区中 16 个最高地址位字节的逻辑 1 的个数没有变化；
- 当固件升级结束时，在 NVM 区中 16 个最高地址位字节的其他位写 0；
- 现在系统可以重新启动，将从非保护区开始运行新的固件。

2.4　随机存储器 RAM

2.4.1　随机存储器 RAM 结构与功能

nRF24LE1 包含两个单独的随机存储器 RAM 块，这些块可用来存储数据或程序。MCU 内部 RAM(IRAM) 的使用是最快和最灵活的，但只有 256 字节。

为了提供更多的临时存储数据或程序代码空间，nRF24LE1 提供一个附加的 1 024×8 位 (1 KB) SRAM 存储器块，默认位于 XDATA 地址空间的 0x0000～0x03FF。SRAM 块在 MCU 的地址空间可以改变。

nRF24LE1 SRAM 块的特殊功能是由两个物理独立的 512 字节块组成的，分别称为数据保持块(低512字节)和数据非保持块。数据保持块在存储器维持低功耗模式时可以保持存储器内容。

每个SRAM块的位置空间是可以配置的,如图2.4.1所示。

可以将SRAM存储器块设置为数据存储器或程序代码存储器地址,由MEMCON寄存器来进行控制,寄存器各位功能如表2.4.1所列。

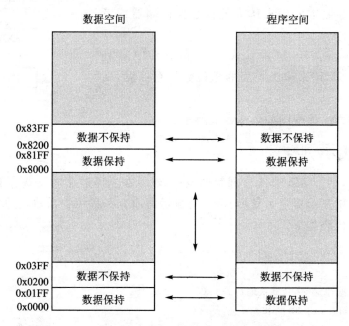

图2.4.1 可配置的SRAM地址空间位置

表2.4.1 MEMCON寄存器各位功能

地址	位	类型	功能	复位值
0xA7	[7:3]		保留	0x00
	[2]	R/W	SRAM地址定位: 0—SRAM块起始地址0x0000 1—SRAM块起始地址0x8000	
	[1]	R/W	数据非保持块映射: 0—映射作为数据存储器 1—映射作为程序存储器	
	[0]	R/W	数据保持块映射: 0—映射作为数据存储器 1—映射作为程序存储器	

2.4.2 SRAM 示例程序流程图

nRF24LE1 提供一个附加的 1 024×8 位(1 KB)SRAM 存储器块,默认位于 XDATA 地址空间的 0x0000~0x03FF。SRAM 的存储大小是 1 024 字节,或者 512 个字。使用时可将这 1 024 字节的存储空间用来作为随机数据存储器或者代码的存储。

本示例程序使用了 SRAM 的默认配置,整个 SRAM 处于 XDATA 地址空间,用来作为扩展的随机数据存储器,而非用于代码的存储。

本示例程序的程序流程图如图 2.4.2 所示。

图 2.4.2 SRAM 示例程序流程图

2.4.3 SRAM 示例程序

在示例程序中,整个 XDATA 全部存储数据,然后在主函数中将这些存储的数据读出来,通过串口送到计算机显示器上去显示,可以对显示的内容和程序中初始化的数据进行比较。

示例程序的源代码如下:

```
/******************************************************
/               **程序说明**
/SRAM 配置成数据存储空间,默认地址是 0x0000~0x03FF
/下面的程序将对这个 1 KB 的 XDATA 进行测试
******************************************************/
#include "reg24le1.h"
/******************************************************
/建立一个全局数组,可以存储 500 个 int 类型的变量
******************************************************/
int xdata dat[500] = {0x00};
/******************************************************
/函数名称:delay()
/函数功能:根据输入参数的大小控制软件延时的长度
/输入参数:x 为延时的时间数
/返回参数:无
******************************************************/
void delay(unsigned int x)
{
    unsigned char di;
```

```c
    for(;x>0;x--)
       for(di = 175;di>0;di--)
         {
           ;
         }
}
```

/**
/函数名称：IO_cof()
/函数功能：初始化 nRF24LE1 的 I/O 口和工作时钟
/输入参数：无
/返回参数：无
***/

```c
void IO_cof()
{
CLKCTRL = 0x28;
CLKLFCTRL = 0x01;

P0CON = 0x00;
P0DIR = 0x00;

P1DIR = 0x00;
P1CON = 0x00;
LED = 0;
}
```

/**
/函数名称：uart_init()
/函数功能：nRF24LE1 的串口初始化
/输入参数：无
/返回参数：无
***/

```c
void uart_init()
{
    P0DIR &= 0xF7;              /* 配置 P0.3(TXD)为输出 */
    P0DIR |= 0x10;              /* 配置 P0.4(RXD)为输入 */
    P0   |= 0x18;
    S0CON = 0x50;
    PCON |= 0x80;               /* 配置波特率倍增 */
```

```
    WDCON |= 0x80;              /*选择使用内部波特率发生器*/
    S0RELL = 0xF3;              /*设置波特率为 38 400*/
    S0RELH = 0x03;
}
/*******************************************************************
/函数名称：send()
/函数功能：串口发送一个字符
/输入参数：ch 为发送的字符
/返回参数：无
*******************************************************************/
void send(char ch)
{
    S0BUF = ch;
    while(!TI0);
    TI0 = 0;
}
/*******************************************************************
/函数名称：puts()
/函数功能：串口发送一个字符串
/输入参数：s 为指向想发送的字符串的指针
/返回参数：无
*******************************************************************/
void puts(unsigned char * s)
{
    while( * s != '\0')
    {
        send( * s++);
    }
}
/*******************************************************************
/函数名称：datainit()
/函数功能：初始化数组 dat 的 500 个数组成员
/输入参数：无
/返回参数：无
*******************************************************************/
void datainit(void)
```

```
{
    int inum = 0;
    for(inum = 0;inum<500;inum++)
        dat[inum] = inum * 2;
}
/******************************************************************
/主函数
******************************************************************/
void main(void)
{
    int num;
    IO_cof();
    uart_init();
    datainit();
    delay(100);
    puts("**********************This Is A nRF24LE1 \
Ram Test Program********************");
    send('\n');
    delay(100);
    for(num = 0;num<500;num++)              /*利用循环在串口上显示dat的成员数据*/
    {
        if((num % 50) == 0)
        {
            send('\n');
        }
        send(((dat[num]/100) + '0'));       /*显示dat数组成员的百位*/
        delay(1);
        send(((dat[num]%100)/10 + '0'));    /*显示dat数组成员的千位*/
        delay(1);
        send((dat[num]%10) + '0');          /*显示dat数组成员的个位*/
        delay(1);
        puts("  ");                         /*显示每个dat数组成员之间的间隔*/
    }
    while(1);                               /*程序进入等待*/
}
```

程序运行后,在计算机显示器上显示的内容如图 2.4.3 所示。

第 2 章　nRF24LE1 的 MCU 与应用

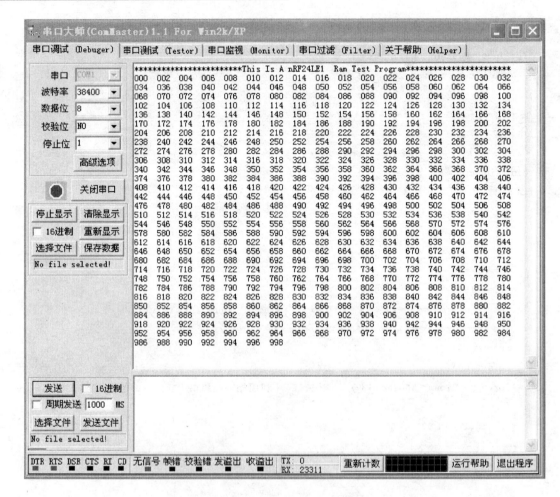

图 2.4.3　XDATA 测试结果

2.5　定时器/计数器

2.5.1　定时器/计数器结构与特性

nRF24LE1 芯片内部包含有一组定时器/计数器,可用来对系统重要事件进行定时。其中一个定时器 RTC2 在低功耗模式下仍可工作,可用来作为唤醒源。

nRF24LE1 包含一组定时器/计数器:

- 3 个 16 位定时器/计数器(Timer0、Timer1 和 Timer2),既可作为基于 MCU 时钟驱动的定时器,也可作为由外部可编程 I/O 信号所驱动的计数器。

- RTC2 是一个具有捕获及比较能力的可配置的、线性的 16 位实时时钟。输入的时钟源频率为 32.768 kHz。

定时器/计数器内部结构方框图如图 2.5.1 所示。

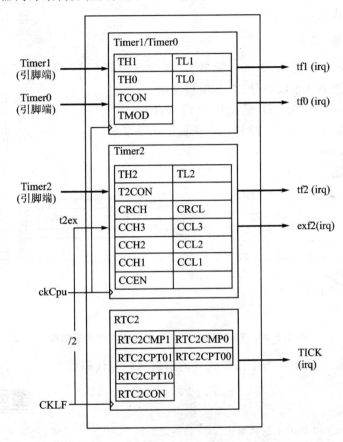

图 2.5.1 定时器/计数器内部结构方框图

2.5.2 Timer0 和 Timer1 的功能与初始化

在定时器模式，Timer0 和 Timer1 每隔 12 个时钟周期递增加 1。

在计数器模式，当检测到引脚端 T0(对应于 Timer0)/引脚端 T1(对应于 Timer1)有下降沿时，计数器加 1。

注意：定时器输入引脚端 T0、T1 和 T2 必须按要求进行配置。

因为需要两个周期来识别 1 到 0 事件发生，最大输入的计数速度为振荡器频率的 1/2。对输入信号的占空比没有要求，但为了确保识别 0 到 1 状态，输入信号必须保持稳定至少一个时钟周期。

第 2 章 nRF24LE1 的 MCU 与应用

Timer0 和 Timer1 的状态和控制在 TCON 和 TMOD 寄存器中完成。16 位定时器的值放在 TH0(高 8 位)和 TL0(低 8 位)。Timer1 与 Timer0 类似,使用 TH1 和 TL1。

Timer0 和 Timer1 可以选择 4 种工作模式。两个特殊功能寄存器 TCON 和 TMOD 用来选择适当的工作模式。

1. 模式 0 和模式 1

在模式 0,Timer0 和 Timer1 每个配置为 13 位寄存器(TL0/TL1＝5 位,TH0/TH1＝8 位)。TL0 和 TL1 的高 3 位没有使用,其值被忽略。

在模式 1,Timer0 和 Timer1 配置为 16 位寄存器。

Timer0 和 Timer1 在模式 0 结构示意图如图 2.5.2 所示。

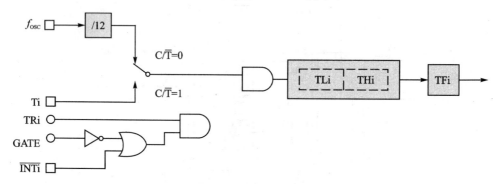

图 2.5.2　Timer0 和 Timer1 在模式 0 结构示意图

Timer0 和 Timer1 初始化为工作模式 0 的示例程序源代码如下:

```
/******************************************************************
/函数名称：timerinit()
/函数功能：Timer0 和 Timer1 的初始化为 13 位的定时器
/输入参数：无
/返回参数：无
******************************************************************/
void timerinit()
{
    TMOD = 0x00;              /*Timer0 和 Timer1 模式控制寄存器,选择定时器工作模式 0*/
    TH0 = (8192 - 8000)/32;   /*预装的数是 8 000*/
    TL0 = (8192 - 8000)%32;
    ET0 = 1;                  /*开启 Timer0 中断*/
    TR0 = 1;                  /*启动 Timer0*/
    TH1 = (8192 - 8000)/32;
    TL1 = (8192 - 8000)%32;
```

        ET1 = 1;
        TR1 = 1;
}

2. 模式 2

在模式 2,Timer0 和 Timer1 均配置为可自动重载的 8 位寄存器。Timer0 和 Timer1 在模式 2 结构示意图如图 2.5.3 所示。

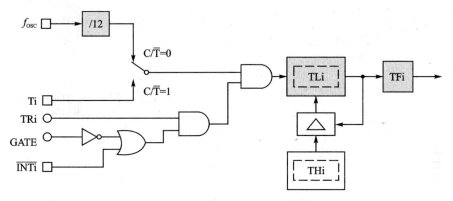

图 2.5.3　Timer0 和 Timer1 在模式 2 结构示意图

Timer0 和 Timer1 配置成工作模式 2,作为 8 位自动重装定时器的示例程序源代码如下:

```
/******************************************************************
/函数名称:timerinit()
/函数功能:Timer0 和 Timer1 的初始化为 8 位自动重装的定时器
/输入参数:无
/返回参数:无
******************************************************************/
void timerinit()
{
    TMOD = 0x22;      /*配置 Timer0 和 Timer1 工作在定时器模式2,自动重装的8位定时器*/
    TH0 = 150;        /*设置自动重装的值为 150*/
    TL0 = 150;        /*设置自动重装的初始值为 150*/
    ET0 = 1;          /*启动 Timer0 中断*/
    TR0 = 1;          /*启动 Timer0 计数*/
    TH1 = 150;
    TL1 = 150;
    ET1 = 1;
    TR1 = 1;
```

3. 模式 3

在模式 3，Timer0 和 Timer1 配置为一个 8 位定时器/计数器和一个 8 位定时器，但是 Timer1 在此模式下保持它的计数。当 Timer0 工作在模式 3 时，Timer1 可以工作在其他模式，如串口的波特率发生器，或一个不需要 Timer1 中断的应用。

Timer0 在模式 3 结构示意图如图 2.5.4 所示。

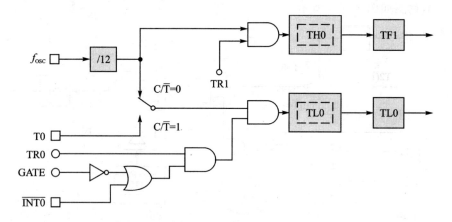

图 2.5.4　Timer0 在模式 3 结构示意图

Timer0 工作在模式 3 时的初始化程序源代码如下：

```
/******************************************************************
/函数名：timerinit()
/函数功能：初始化 Timer0 为工作模式 3,8 位的定时器
/输入参数：无
/返回参数：无
******************************************************************/
void timerinit()
{
    TMOD = 0x03;        /*配置 Timer0 工作模式 3*/
    TH0 = 100;          /*Timer1 预装值为 100*/
    TL0 = 100;          /*Timer0 预装值为 100*/
    ET1 = 1;            /*启动 Timer1 中断*/
    ET0 = 1;            /*启动 Timer0 中断*/
    TR1 = 1;            /*启动 Timer1*/
    TR0 = 1;            /*启动 Timer0*/
}
```

2.5.3 Timer2 的功能与初始化

Timer2 由 T2CON 控制,数值放在 TH2 和 TL2 中。Timer2 有 4 个捕获和 1 个捕获/重载寄存器,当 Timer2 达到 0 时,可以读出值而不需要暂停或重载 16 位的值。

Timer2 可以作为一个定时器、事件计数器或门控定时器。

1. Timer2 重载模式

Timer2 重载模式如图 2.5.5 所示。

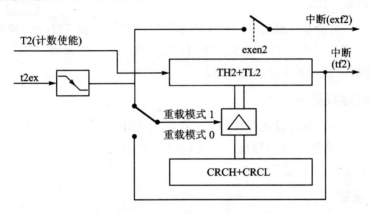

图 2.5.5 Timer2 重载模式

Timer2 重载的初始化程序源代码如下:

```
/******************************************************************
/函数名称:timer2init()
/函数功能:初始化 Timer2 成自动重载模式
/输入参数:无
/返回参数:无
******************************************************************/
void timer2init(void)
{
TH2 = (65536 - 50000)/256;
TL0 = (65536 - 50000)%256;
CRCH = (65536 - 50000)/256;
CRCL = (65536 - 50000)%256;
T2PS = 1;                    /*设置 Timer2 为重载模式 0,时钟为设置为 24 分频*/
T2I0 = 1;
ET2 = 1;
T2R1 = 1;                    /*启动 Timer2*/
T2R0 = 0;
}
```

2. 定时器模式

定时器模式可以通过设置 T2CON 寄存器中的 T2I0＝1 和 T2I1＝0 位来调用。在此模式式,计数器的计数源来自于时钟输入。

Timer2 每 12 或 24 个时钟周期即加 1(取决于 2∶1 预分频器)。预分频器由 T2CON 寄存器的 T2PS 位确定。当 T2PS＝0 时,每隔 12 个时钟周期 Timer2 加 1;否则是 24 个时钟周期加 1。

3. 事件计数器模式

事件计数器模式可以通过设定 T2CON 寄存器的 T2I0＝0 和 T2I1＝1 位来调用。在此模式,当 T2 引脚端外部信号由 1 到 0 发生改变时,Timer2 加 1。引脚端 T2 输入在时钟的上升沿被采样,最大计数速度为系统时钟的 1/2。

4. 门控定时器模式

门控定时器模式可以通过设定 T2CON 寄存器的 T2I0＝1 和 T2I1＝1 位来调用。在此模式,每 12 或 24 个时钟周期即加 1(取决于 T2CON 的 T2PS 标志),并且受 T2 引脚端外部信号控制,当 T2＝0 时,Timer2 停止工作。

5. Timer2 重载

从 CRC(Compare/Reload Capture)寄存器来的 16 位重载可用以下两种模式来实现:
- 重载模式 0——重载信号由 Timer 2 溢出来产生(自动重载);
- 重载模式 1——重载信号由 t2ex 负跳变时产生。

注意:t2ex 连接的内部时钟信号是 CLKLF 的一半。

2.5.4 定时器/计数器的特殊功能寄存器 SFR

定时器/计数器的特殊功能寄存器 SFR 有:

(1) 定时器/计数器控制寄存器 TCON

TCON 寄存器反映 MCU Timer0 和 Timer1 的状态,并用来控制其工作模式。

当相应的服务程序被调用后,tf0、tf1(Timer0 和 Timer1 溢出标志),ie0 和 ie1(外部中断 0 和 1 标志)均由硬件自动清除(清 0)。

(2) 定时器模式寄存器 TMOD

TMOD 寄存器用来配置 Timer0 和 Timer1。

(3) Timer0(TH0、TL0)

TH0 和 TL0 寄存器反映 Timer0 的状态,TH0 保存高字节,TL0 保存低字节。Timer0 可以配置成计数器或定时器。

(4) Timer1(TH1、TL1)

TH1 和 TL1 寄存器反映 Timer1 的状态，TH1 保存高字节，TL1 保存低字节。Timer1 可以配置成计数器或定时器。

(5) Timer2 控制寄存器 T2CON

T2CON 寄存器反映 Timer2 当前的状态，并控制 Timer2 的工作模式。

(6) Timer2(TH2、TL2)

TL2 和 TH2 寄存器反映 Timer2 的状态，TH2 保存高字节，TL2 保存低字节。Timer2 可以配置成比较、捕获或重载模式。

(7) 比较/捕获使能寄存器 CCEN

比较/捕获单元与 Timer2 相关联，CCEN 是其配置寄存器。

(8) 捕获寄存器 CC1、CC2 和 CC3

比较/捕获寄存器(CC1、CC2 和 CC3)都是 16 位寄存器，与 Timer2 相关联，CCHn 保存高字节，CCLn 保存低字节，$n=1、2、3$。

(9) 比较/重载/捕获寄存器 CRCH、CRCL

CRC（Compare/Reload/Capture）寄存器是一个与 Timer2 相关联的 16 位寄存器。CRCH 保存高字节，CRCL 保存低字节。

有关定时器/计数器的特殊功能寄存器 SFR 的各位定义与功能的更多内容请登录 www.nordicsemi.com，查询 nRF24LE1 Ultra-low Power Wireless System On-Chip Solution Preliminary Product Specification v1.6。

2.5.5 实时时钟 RTC

RTC2 是从 0 开始向上以 32.768 kHz 的时钟频率计数的一个 16 位定时器。RTC2 包含两个寄存器，用来捕获定时器的值：一个由 32.768 kHz 时钟的下降沿装入；另一个由 CPU 的时钟驱动，以获得更高的分辨率。两个寄存器均可由外部事件的结果来更新。当定时器和比较器寄存器的值相同时，RTC2 产生一个中断，RTC2 确保唤醒的功能优于中断。

RTC 时钟频率为 32.768 kHz，电流消耗为 μA 级，与中断(TICK)比较分辨率为 30.52 μs，增强型捕获的分辨率为 125 ns。

RTC 由 RTC 特殊寄存器控制，有关 RTC 特殊寄存器的各位定义与功能的更多内容请登录 www.nordicsemi.com，查询 nRF24LE1 Ultra-low Power Wireless System On-Chip Solution Preliminary Product Specification v1.6。

2.5.6 定时器/计数器示例程序流程图

在本示例程序中，nRF24LE1 的 P0.7 是 T0 定时器/计数器的输入端，nRF24LE1 的 P0.5 作为脉冲信号的发生端。示例程序的流程图如图 2.5.6 所示。

第 2 章 nRF24LE1 的 MCU 与应用

在本示例程序运行后,nRF24LE1 的 P0.5 引脚端会有规律的脉冲信号输出,通过杜邦线将其连接到 nRF24LE1 的 P0.7 引脚端上,nRF24LE1 的定时器/计数器对 P0.5 上的脉冲的下降沿进行计数,并不断地将当前记录的数据通过串口发送给计算机,在计算机的显示器上可以看见当前的记录数。

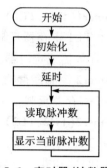

图 2.5.6 定时器/计数器示例程序的流程图

2.5.7 定时器/计数器示例程序

定时器/计数器示例程序的源代码如下:

```
#include "def.h"
#include "reg24le1.h"
/*****************************************************
/函数名称:delay()
/函数功能:根据参数 x 实现软件延时功能
/输入参数:x 为延时时间数
/返回参数:无
*****************************************************/
void delay(unsigned int x)
{
unsigned char j = 120;
    for(;x>0;x--)
      for(;j>0;j--)
          {
            ;
          }
}
/*****************************************************
/函数名称:init_counter()
/函数功能:初始化计数器
/输入参数:无
/返回参数:无
*****************************************************/
void init_counter(void)
{
    TMOD = 0x05;      /*配置计数器 0 成沿计数方式*/
    TR0 = 1;          /*启动计数器*/
    TH0 = 0;          /*高位清 0*/
    TL0 = 0;          /*低位清 0*/
```

```
}
/*************************************************************/
/函数名称: init_uart()
/函数功能: 初始化 nRF24LE1 的异步串口
/输入参数: 无
/返回参数: 无
*************************************************************/
void init_uart(void)
{
    CLKCTRL = 0x28;            /* 设置 nRF24LE1 工作时钟采用 XCOSC 16 MHz */
    CLKLFCTRL = 0x01;
    P0DIR &= 0xF7;             /* 配置 P0.3(TXD)为输出 */
    P0DIR |= 0x10;             /* 配置 P0.4(RXD)为输入 */
    P0 |= 0x18;
    S0CON = 0x50;
    PCON |= 0x80;              /* 设置波特率倍增 */
    WDCON = 0x80;              /* 选用内部波特率发生器 */
    S0RELL = 0xFB;
    S0RELL = 0xF3;             /* 设置波特率为 38 400 */
}
/*************************************************************/
/函数名称: io_config()
/函数功能: 初始化 nRF24LE1 的 I/O 口
*************************************************************/
void io_config(void)
{
    P0DIR &= ~(0x20);          /* 配置 P0.5 为输出作为脉冲发生端口 */
    P0DIR |= 0x80;             /* 配置 P0.7 为输入,P0.7 是一个复用口,这里是作为 T0 */
    P05 = 0;
}
/*************************************************************/
/计数器脉冲读取的函数,返回的是当前计数器的脉冲数目
/函数名称: counter_result()
/输入参数: 无
*************************************************************/
unsigned int counter_result(void)
{
    static unsigned int flowtime = 0;    /* 记录脉冲超过 50 000 个的次数 */
    unsigned int res;                    /* 脉冲数保存 */
    res = TH0 * 256 + TL0;               /* 计算当前记录的脉冲个数 */
    if(res == 50000)                     /* 脉冲累积到 50 000 个就清零计数器 */
```

第 2 章　nRF24LE1 的 MCU 与应用

```
    {
    flowtime ++ ;
    res = 0;
    }
    return (flowtime * 50000 + res);    /* 返回当前记录的脉冲总数 */
    }
/*******************************************************************
/通过串口打印一个字符到计算机的串行终端
 *******************************************************************/
void putch(char s)
{


SOBUF = s;
while(!TI0);                            /* 等待发送完成 */
TI0 = 0;
}
/*******************************************************************
/主函数部分
 *******************************************************************/
unsigned char num[5];                   /* 定义一个数组来保存脉冲数,最多 5 位数 */
void main(void)
{
unsigned int result;                    /* 保存脉冲的个数 */
io_config();                            /* I/O 口的配置函数 */
init_counter();                         /* 初始化计数器 */
init_uart();                            /* 初始化串口 */
while(1)
 {
 delay(60000);                          /* 软件延时,这里延时时间较长 */
 P05 = !P05;                            /* P0.5 配置成计数器的脉冲输入信号 */
 delay(60000);                          /* 软件延时,这里延时时间较长 */
 delay(60000);                          /* 软件延时,这里延时时间较长 */
 delay(60000);                          /* 软件延时,这里延时时间较长 */
 delay(60000);                          /* 软件延时,这里延时时间较长 */
 P05 = !P05;                            /* P0.5 配置成计数器的脉冲输入信号 */
 result = counter_result();             /* 读取计数的脉冲结果 */
/*******************************************************************
/分离脉冲个数的每一位,然后利用串口进行显示
 *******************************************************************/
num[0] = result/10000;                  /* 获得万位的数据 */
```

```c
    putch(TOASC(num[0]));
num[1] = (result % 10000)/1000;      /*获得千位的数据*/
    putch(TOASC(num[1]));
num[2] = (result % 1000)/100;        /*获得百位的数据*/
    putch(TOASC(num[2]));
num[3] = (result % 100)/10;          /*获得十位的数据*/
    putch(TOASC(num[3]));
num[4] = (result % 10);              /*获得个位的数据*/
    putch(TOASC(num[4]));
    putch('\n');                     /*换行*/
    }
}
```

程序运行结果:

nRF24LE1通过串口线连接到计算机上,程序下载到nRF24LE1,复位后,每隔1个时间段,数据就会加1。在计算机的显示器上可以看见当前的记录数,如图2.5.7所示。

图2.5.7 计数器串口脉冲数显示截图

2.6 中断

2.6.1 中断源和中断向量

nRF24LE1 有一个具有 18 个中断源的中断控制器。中断控制器管理基于动态编程次序的实时事件的信号中断，如定时器、射频收发器、引脚端有效，等等。

nRF24LE1 的中断具有 8 个中断源 4 个优先级，可用的中断请求标志以及可选择极性的引脚端中断。

当一个被使能的中断发生时，MCU 将执行该中断向量地址所指向的中断服务程序，如表 2.6.1 所列，除非有更高优先级的中断发生，MCU 将一直执行完成该中断服务程序。

表 2.6.1 nRF24LE1 中断源和中断向量地址

中断源	中断向量	极 性	描 述
IFP	0x0003	低电平/下降沿	来自引脚端的中断
tf0	0x000B	高电平	Timer0 溢出中断
POFIRQ	0x0013	低电平/下降沿	电源故障中断
tf1	0x001B	高电平	Timer1 溢出中断
ri0	0x0023	高电平	串行通道接收中断
ti0	0x0023	高电平	串行通道发射中断
tf2	0x002B	高电平	Timer2 溢出中断
exf2	0x002B	高电平	Timer2 外部重载
RFRDY	0x0043	高电平	RF SPI 就绪
RFIRQ	0x004B	下降沿/上升沿	RF 中断
MSDONE	0x0053	下降沿/上升沿	主 SPI 传输完成中断
WIRE2IRQ	0x0053	下降沿/上升沿	2 线传输完成中断
SSDONE	0x0053	下降沿/上升沿	从 SPI 传输完成中断
WUOPIRQ	0x005B	上升沿	引脚端唤醒中断
MISCIRQ	0x0063	上升沿	杂项中断： XOSC16M 启动中断(X16IRQ) ADC 就绪中断(ADCIRQ) RNG 就绪中断(RNGIRQ)
TICK	0x006B	上升沿	内部唤醒中断(来自 RTC2)

注意：

① 当 XOSC16M 已经启动时，X16IRQ 阻塞 ADC 和 RNG 的中断控制，因此推荐通过清除 CLKCTRL[3] 来禁止 X16IRQ 中断，而 XOSC16M 启动仍然可以查询。

② 除非唤醒已通过 WUCON 使能，否则 RFIRQ、WUOPIRQ、MISCIRQ 和 TICK 将不会激活。

2.6.2 中断用特殊功能寄存器 SFR

中断控制使用到 TCON、T2CON、IRCON、SCON、IP0、IP1、IEN0、IEN1 和 INTEXP 特殊功能寄存器。不同的 SFR 寄存器用来控制不同中断的优先次序。

(1) 中断使能寄存器 0 IEN0

IEN0 寄存器控制全局中断的使能/禁止，包括 Timer0、Timer1、Timer2 以及通道 0 和串口中断的使能/禁止。

(2) 中断使能寄存器 1 IEN1

IEN1 寄存器控制 RF、SPI 和 Timer2 中断。2 线主 SPI 和从 SPI 使用同一个中断线。

(3) 中断优先级寄存器 IP0 和 IP1

14 个中断源可以分成 6 组，每一组可以选择 4 个优先级之一，可以通过设置 IP0 和 IP1 寄存器的相应值来实现。

(4) 中断请求控制寄存器 IRCON

IRCON 包括 Timer2、SPI、RF、USB 和唤醒中断请求标志。

有关中断用特殊功能寄存器的更多内容请登录 www.nordicsemi.com，查询 nRF24LE1 Ultra-low Power Wireless System On-Chip Solution Preliminary Product Specification v1.6。

2.6.3 中断示例外接电路

在本示例中，需要用到外部中断和定时器中断两种中断，通过外部中断来控制定时器中断，默认情况下外部中断的优先级是最高的。外部中断的触发按键电路连接图如图 2.6.1 所示。nRF24LE1 的 I/O 引脚 P0.5 作为 INT0 输入。

在本示例中还需要使用一个蜂鸣器，定时器中断时，使蜂鸣器发出间断性的声响。蜂鸣器电路如图 2.6.2 所示。nRF24LE1 的 I/O 引脚 P1.2 作为蜂鸣器驱动输出。

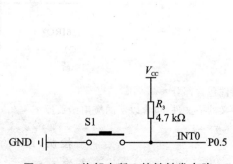

图 2.6.1 外部中断 0 按键触发电路

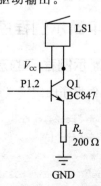

图 2.6.2 蜂鸣器电路

2.6.4 中断示例程序流程图

本示例程序的程序流程图如图 2.6.3 所示。

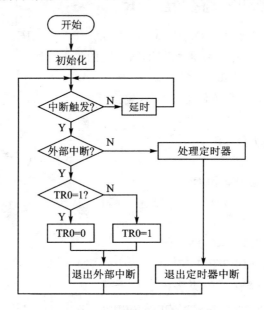

图 2.6.3 中断示例程序流程图

本示例程序实现的功能是通过外部中断来控制定时器中断。定时器中断在打开时,根据设定的中断时间,连续发生中断。所设定的定时器 0 中断时间间隔是 0.025 s,中断 40 次就是 1 s。在程序中,定时器中断每隔 1 s 使 nRF24LE1 的 P1.2 引脚端的电平翻转一次,使与之连接的蜂鸣器周期性地发出声音,叫 1 s,停 1 s,一直这样下去。

此时,如果按下 S1,那么定时器中断会被关闭,蜂鸣器保持最后一次的状态。如果再一次按下 S1,则定时器中断会打开,恢复之前的状态。

2.6.5 中断示例程序

1. 中断示例程序的宏定义

中断示例程序的宏定义如下:

```
#include "reg24le1.h"          /*头文件包含部分*/
#include "inc.h"               /*自定义的头文件,包含以下的应用函数原型*/
/*************************************************************/
#define ONESTEP    33333
#define BUZZER     P12
```

```
/*************************************************************
/函数名称：delay()
/函数功能：实现软件延时
/输入参数：dj 为软件延时的时间数
/返回参数：无
*************************************************************/
void delay(unsigned int dj)
{
unsigned char di;
 for(;dj>0;dj--)
    for(di=120;di>0;di--)
    {
     ;
    }
}
```

2. 中断示例程序的初始化函数

中断示例程序的初始化函数如下：

```
/*************************************************************
/函数名称：ioconfig()
/函数功能：初始化 nRF24LE1 的 I/O 口
/输入参数：无
/返回参数：无
*************************************************************/
void ioconfig()
{
P1DIR &= 0xFB;                  /* 配置 P1.2 为输出 */
P12 = 0;
P0DIR |= 0x20;                  /* 配置 P0.5 为输入 */
P05 = 1;
}
/*************************************************************
/函数名称：uart()
/函数功能：初始化 nRF24LE1 的串行口
/输入参数：无
/返回参数：无
*************************************************************/
void uart()
{
```

```c
        CLKCTRL = 0x28;                 /* 设置 MCU 时钟为 16 MHz */
        CLKLFCTRL = 0x01;               /* 设置实时时钟为 32.768 kHz */
        P0DIR &= 0xF7;                  /* 配置 P0.3(TXD) 为输出 */
        P0DIR |= 0x10;                  /* 配置 P0.4(RXD) 为输入 */
        P0 |= 0x18;
        S0CON = 0x50;
        PCON |= 0x80;                   /* 波特率倍增 */
        WDCON |= 0x80;                  /* 选择内部波特率发生器 */
        S0RELL = 0xFB;                  /* 波特率为 38 400 */
        S0RELL = 0xF3;
}
/***********************************************************
/函数名称：int0cof()
/函数功能：外部中断 0 初始化
/输入参数：无
/返回参数：无
***********************************************************/
void int0cof()                          /* int0 中断寄存器配置 */
{
    INTEXP = 0x08;                      /* 外部中断 0 引脚为 P0.5 */
    TCON |= 0x01;                       /* 下降沿触发 */
    IEN0 |= 0x01;                       /* 外部中断使能 */
}
/***********************************************************
/函数名称：timer0cof()
/函数功能：Timer0 初始化
/输入参数：无
/返回参数：无
***********************************************************/
void timer0cof()                        /* Timer0 中断配置 */
{
    TMOD = 0x01;                        /* 定时器模式 1 */
    TH0 = (65536 - ONESTEP)/256;        /* 定时器初始值装载 */
    TL0 = (65536 - ONESTEP)%256;
    ET0 = 1;                            /* 打开 Timer0 中断允许 */
    TR0 = 1;                            /* 启动 Timer0 */
}
```

3. 中断示例程序的中断服务函数

中断示例程序的中断服务函数如下：

```
/***************************************************************
/函数名称：ex0service()
/函数功能：处理外部中断
/输入参数：无
/返回参数：无
***************************************************************/
void ex0service() interrupt INTERRUPT_IFP        /*外部中断服务函数*/
{
if(TR0)
TR0 = 0;
else
TR0 = 1;
puts("one pin interrupt toggled!");              /*进入外部中断时的调试信息*/
putch('\n');                                     /*换行*/
}
/***************************************************************
/函数名称：time0service()
/函数功能：处理Timer0中断
/输入参数：无
/返回参数：无
***************************************************************/
void time0service() interrupt 1
{
static char num = 0;
TH0 = (65536 - ONESTEP)/256;                     /*重新装载计数次数*/
TL0 = (65536 - ONESTEP) % 256;
num ++ ;
if(num == 40)                                    /*1s取反一次，即1s有一次电平翻转*/
{
num = 0;
BUZZER = !BUZZER;                                /*连接蜂鸣器引脚端的电平翻转*/
}
}
```

4. 中断示例程序的串口功能函数

中断示例程序的串口功能函数如下：

```
/*************************************************************
/函数名称：putch()
/函数功能：串口发送一个字符
/输入参数：ch 为待发送的字符
/返回参数：无
**************************************************************/
void putch(char ch)
{
S0BUF = ch;                           /*写到发送的缓冲*/
while(!TI0);                          /*等待发送完成*/
TI0 = 0;                              /*清除发送完成标志*/
}
/*************************************************************
/函数名称：puts()
/函数功能：串口发送一个字符串
/输入参数：str 为字符串指针
/返回参数：无
**************************************************************/
void puts(char * str)
{
while( * str != '\0')                 /*字符串没有结束*/
{
putch( * str ++ );                    /*发送字符*/
}
}
```

5. 中断示例程序的主函数

中断示例程序的主函数如下：

```
/*************************************************************
/主函数部分
**************************************************************/
void main()
{
EA = 0;                               /*关中断*/
uart();                               /*串口配置*/
int0cof();                            /*外部中断配置*/
timer0cof();                          /*定时器中断配置*/
ioconfig();                           /*I/O 口配置*/
```

```
puts("this is a interrupt test program!");   /*打印显示提示语句*/
putch('\n');
EA = 1;                                       /*开中断*/
    while(1)
    {
    delay(10);
    }

}
```

程序执行结果：

程序编译通过后，将生成的 HEX 文件下载到 nRF24LE1 中，然后复位执行程序，会得到上述的功能。

2.7 看门狗

2.7.1 看门狗结构与功能

当软件因为某种原因挂起时，片上看门狗计数器能够强制系统复位。

看门狗的时钟频率为 32.768 kHz，电流消耗为 μA 级，最小的看门狗超时间隔为 7.812 5 ms，最大的看门狗超时间隔为 512 s，禁用（复位）只能由系统复位。

看门狗内部结构方框图如图 2.7.1 所示。

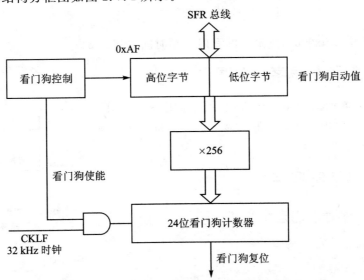

图 2.7.1　看门狗内部结构方框图

2.7.2 看门狗寄存器 WDSV

WDSV 寄存器用来控制看门狗。看门狗寄存器功能如表 2.7.1 所列。

表 2.7.1 看门狗寄存器功能

地址	名称	位	复位值	类型	功能
0xAF	WDSV	[15:0]	0x0000	R/W	看门狗寄存器的初始值。 包含看门狗定时器的初始值。如果寄存器装入 0x0000，则最大的看门狗超时间隔被使用，看门狗没有被禁用。LSB 总是被先写或先读，然后到最高位。 在 LSB/MSB 的写和读的指针是分开的。在读访问后，写指针总是指向 LSB 的。经过写访问后，读指针总是指向 LSB 的

复位后，看门狗的默认状态是禁止。当 WDSV 的两个字节都被写入后，看门狗将有效，首先写入 LSB。看门狗计数器向下计数（倒计数），当到达 0 时，整个微控制器包括外设将被复位。

看门狗必须写入一个 WDSV 才能够重新启动。当写入新的启动值时，看门狗计数器将更新并重新开始计数。为避免复位，软件必须确保能够重载 WDSV。看门狗一旦启动，只能由系统复位或芯片进入节电模式来禁止。

看门狗计数器使用 32.768 kHz 时钟，当使用看门狗时，32.768 kHz 时钟源必须使能。
看门狗超时间隔由下式确定（单位为 s）：

$$看门狗超时间隔 = \frac{WDSV \times 256}{32\,768}$$

如果 WDSV 写入 0x0000，则最大的看门狗超时间隔将为 512 s。

2.7.3 看门狗示例程序流程图

每隔一段时间本示例程序能够通过看门狗使 nRF24LE1 复位，同时通过计算机串口在显示器上显示提示信息。本示例程序的流程图如图 2.7.2 所示。

2.7.4 看门狗示例程序

1. 看门狗示例程序的宏定义

看门狗示例程序的宏定义如下：

```
#include "reg24le1.h"
#include "intrins.h"
```

图 2.7.2 nRF24LE1 看门狗示例程序流程图

```
#define LED P05           /*定义P0.5为LED驱动*/
#define OCRG 0x10
#define RSTP 0x20
#define DOGR 0x30
#define HDEG 0x40;
#define TRUE  1           /*定义常量为真*/
#define FALSE 0           /*定义常量为假*/
/****************************************************************
/函数原型声明部分
 ****************************************************************/
void io_config();
void watchdog_init();
void delay(unsigned int x);
void decount();
void puts(unsigned char * s);
void send(char ch);
void redrst();
void uart_init();
```

2. 看门狗示例程序的初始化函数

看门狗示例程序的初始化函数如下:

```
/****************************************************************
/函数名称: io_config()
/函数功能: nRF24LE1的I/O初始化
/输入参数: 无
/返回参数: 无
 ****************************************************************/
void io_config()
{
CLKCTRL = 0x28;
CLKLFCTRL = 0x01;
P0CON = 0x00;
P0DIR = 0x00;
LED = 0;
}
/****************************************************************
```

```
/*函数名称：uart_init()
/*函数功能：初始化 nRF24LE1 的串口
/*输入参数：无
/*返回参数：无
**************************************************************/
void uart_init()
{
    P0DIR &= 0xF7;              /* 配置 P0.3(TXD)为输出 */
    P0DIR |= 0x10;              /* 配置 P0.4(RXD)为输入 */
    P0   |= 0x18;
    S0CON = 0x50;
    PCON |= 0x80;               /* SMOD = 1 */
    WDCON |= 0x80;              /* 内部波特率发生器 */
    S0RELL = 0xF3;              /* 设置波特率为 38 400 */
    S0RELH = 0x03;
}
/***************************************************************
/*函数名称：watchdog_init()
/*函数功能：初始化看门狗寄存器
/*输入参数：无
/*返回参数：无
**************************************************************/
void watchdog_init()
{
WDSV = (640 % 256);
delay(10);
WDSV = (640/256);
puts("\n   the le1 will restart in 5 seconds! \n");
}
```

3. 看门狗示例程序的功能函数

看门狗示例程序的功能函数如下：

```
/***************************************************************
/*函数名称：delay()
/*函数功能：实现软件延时
```

/输入参数：无
/返回参数：无
**/

```c
void delay(unsigned int x)
{
unsigned char di;
    for(;x>0;x--)
        for(di=175;di>0;di--)
            {
                ;
            }
}
```

/**
/函数名称：redrst()
/函数功能：获取上一次复位的原因并在串口上显示出来
/输入参数：无
/返回参数：无
**/

```c
void redrst()
{
unsigned char rst;
rst = RSTREAS;
rst &= 0x07;
send('\n');
switch(rst)
{
  case 0x00:puts("On-chip reset generator!");break;    /*片上晶振产生复位*/
  case 0x01:puts("RST pin!");break;                    /*引脚复位*/
  case 0x02:puts("Watchdog!");break;                   /*看门狗复位*/
  case 0x04:puts("Reset from on-chip hardware debugger!");break;/*硬件调试软件复位*/
  case 0x06:puts("first reset by debugger,then by watchdog!");break;/*首先是DEBUG复位，其次
                                                              是看门狗复位*/
  default:puts("error!");break;                        /*否则就是错误类型*/
}
RSTREAS = 0x00;
}
```

```
/*************************************************************
/函数名称: send()
/函数功能: 串口发送一个字符
/输入参数: ch 为待发送的字符
/返回参数: 无
*************************************************************/
void send(char ch)
{
  S0BUF = ch;
  while(!TI0);
  TI0 = 0;
}
/*************************************************************
/函数名称: puts()
/函数功能: 串口发送一个字符串
/输入参数: s 为字符串的指针
/返回参数: 无
*************************************************************/
void puts(unsigned char * s)
{
  while( * s ! = '\0')
  send( * s++ );
}
/*************************************************************
/函数名称: decount()
/函数功能: 看门狗的状态信息反馈
/输入参数: 无
/返回参数: 无
*************************************************************/
void decount()
{
unsigned int wh,wl;
unsigned int time0 = 5,time1 = 0;
wl = WDSV;
delay(10);
wh = WDSV;
```

```
wl &= 0xFF;
wh &= 0xFF;
time1 = (wh * 256 + wl) * 256/32768;        /* 读出初始时的计数 */
/*
if(time1 != time0)
   {
LED = !LED;
time0 = time1;
send(time0 + '0');
send('\n');
   }
         */
delay(10000);
LED = !LED;
}
```

4. 看门狗示例程序的主函数

看门狗示例程序的主函数如下：

```
/***************************************************************
/主函数部分
***************************************************************/
void main(void)
{
io_config();                /* I/O 配置函数 */
uart_init();                /* 串口初始化函数 */
redrst();                   /* 读取和显示最近一次复位的方式 */
watchdog_init();            /* 看门狗初始化 */
while(1)
   {
decount();                  /* 等待复位 */
   }
}
```

程序运行结果如图 2.7.3 所示。

下载本示例程序到 nRF24LE1 后，按复位键复位 nRF24LE1，通过串口在计算机显示器上显示"RST pin!"，之后显示的全部是"Watchdog!"，每 5 s 看门狗就自动复位一次。

第2章 nRF24LE1 的 MCU 与应用

图 2.7.3 程序运行结果（显示器截图）

2.8 功耗和时钟管理

2.8.1 工作模式

nRF24LE1 的工作模式包含有运行模式，深度睡眠模式，存储器维持、定时器关闭模式，存储器维持、定时器开启模式，寄存器维持、定时器关闭模式，寄存器维持、定时器开启模式，待机模式。有关 nRF24LE1 工作模式的更多内容请登录 www.nordicsemi.com，查询 nRF24LE1 Ultra-low Power Wireless System On-Chip Solution Preliminary Product Specification v1.6。

在复位或上电后，nRF24LE1 进入有效工作模式，功能的实现由软件进行控制。为进入其中一种节电模式，必须按照所选定的工作模式写入相应内容到 PWRDWN 寄存器。若要从节电模式再次进入有效工作模式，则需要一个唤醒源来激活。

2.8.2 功耗和时钟管理有关的寄存器

nRF24LE1 通过管理工作模式以及控制工作的时钟频率来实现功耗管理功能。与功耗和时钟管理有关的寄存器如图 2.8.1 所示。

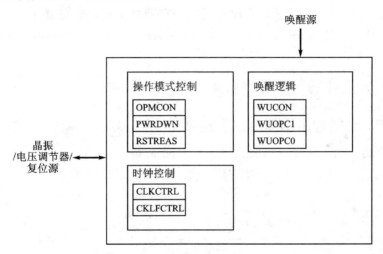

图 2.8.1 与功耗和时钟管理有关的寄存器

MCU 的时钟(CKCPU)由片内 RC 振荡器或晶体振荡器来产生。

微控制器系统的时钟源和频率由 CLKCTRL 寄存器控制。

注意：CLKCTRL 寄存器不支持读/修改/写操作。

32 kHz 时钟源(CLKLF)由 CLKLFCTRL 寄存器控制。

注意：如果选择 CLKLF 作为时钟源，除非 CLKLF 是有效的，否则 MCU 将不会启动。

例如，当选择 CLKLF 源来自外部 I/O 引脚端时，外部时钟源必须是有效的，以确保 MCU 在存储器维持模式下可以被 I/O 唤醒。

节电控制寄存器(PWRDWN)被 MCU 用来控制系统进入节电模式。

工作模式控制寄存器(OPMCON)用来实现控制工作模式下的一些特殊功能。

片内复位电路复位、外部复位引脚复位、看门狗复位和片内硬件调试器复位 4 个复位源可以产生同一个复位和启动顺序。复位结果寄存器(RSTREAS)保留了上一次复位的原因，全部清 0 表示上一次复位是由片内复位电路产生的。对该寄存器的写操作将清除所有位。除非读以后进行清除操作(片内复位或写操作)，RSTREAS 的值将是累加的；也就是说，如果一个调试器的复位后接着一个看门狗复位，RSTREAS 的值将为 110。

在待机模式，RFIRQ、TICK(来自 RTC2)、WUOPIRQ 和 MISCIRQ 唤醒源是可用的，在唤醒配置寄存器(WUCON)中设置。

引脚唤醒功能通过 WUOPC1 和 WUOPC2 两个寄存器来配置。WUOPCx 寄存器的功能

第 2 章 nRF24LE1 的 MCU 与应用

取决于所用的封装类型,对应 nRF24LE1 的不同封装,可唤醒 I/O 引脚端与 WUOPCx 寄存器的对应关系不同。

选择引脚端唤醒高电平有效。如果使能 SPI 的从模式,SPICON0 寄存器的位 0 置位,则 spiSlaveCsn 信号成为一个低电平引脚有效的唤醒源。

有关 nRF24LE1 与功耗和时钟管理有关的寄存器的更多内容请登录 www.nordicsemi.com,查询 nRF24LE1 Ultra-low Power Wireless System On-Chip Solution Preliminary Product Specification v1.6。

2.8.3 功耗和时钟管理示例程序

1. 功耗和时钟管理示例程序的宏定义

下面是一个功耗和时钟管理示例程序。程序完成 nRF24LE1 进入 Deep Sleep 低功耗模式后,利用外部引脚的高电平唤醒 nRF24LE1 进入到正常运行模式。

程序源代码如下:

```
#include "reg24le1.h"      /*头文件包含部分*/
/**************************************************************
/宏定义低功耗模式
***************************************************************/
#define  DeepSleep     0x00
#define  Mem_Ret_On    0x01
#define  Mem_Ret_Off   0x02
#define  Reg_Ret_On    0x03
#define  Reg_Ret_Off   0x04
#define  Standby       0x05
/**************************************************************
/宏定义定时器中断号
***************************************************************/
#define  TF0           0x01
/**************************************************************
/定义布尔逻辑常量
***************************************************************/
#define  true          0x01
#define  false         0x00
/**************************************************************
/定义一个 LED 指示灯
***************************************************************/
#define  LED           P07
```

```
/*中断开关宏定义
*************************************************************/
#define Disableint()    {EA = 0;}
#define Enableint()     {EA = 1;}
/*************************************************************
/定义全局变量 seconds 用于计秒数
*************************************************************/
static unsigned char seconds = 0;
/*************************************************************
/函数原型声明方便其他函数调用
*************************************************************/
void getlastpowerdownmode(void);
/*************************************************************
/函数名称：delay()
/函数功能：延时函数,通过修改延时参数实现延时的不同
/传入函数：x 为延时的时间数
/返回参数：无
*************************************************************/
void delay(int x)
{
int i,j;
 for(i = x;i>0;i--)
    for(j = 120;j>0;j--)
      {
        ;
      }
}
```

2. 功耗和时钟管理示例程序的初始化函数

功耗和时钟管理示例程序的初始化函数如下：

```
/*************************************************************
/函数名称：baudrate()
/函数功能：nRF24LE1 串口的初始化
/输入参数：rate 为设定的波特率
/返回参数：无
*************************************************************/
void baudrate(int rate)
{
    P0DIR &= 0xF7;              /* 配置 P0.3(TXD)为输出 */
```

```c
        P0DIR |= 0x10;              /*配置 P0.4(RXD)为输入*/
        S0CON = 0x50;
        PCON |= 0x80;               /*波特率倍增*/
        WDCON |= 0x80;              /*选用内部波特率发生器*/
    switch(rate)
    {
        case 38400:
                {
                    S0RELL = 0xF3;
                    S0RELH = 0x03;
                }
                    break;
        case 19200:
                {
                    S0RELL = 0xE6;
                    S0RELH = 0x03;
                }
                    break;
        case 14400:
                {
                    S0RELL = 0xDE;
                    S0RELH = 0x03;
                }
                    break;
        case 9600 :
                {
                    S0RELL = 0xCC;
                    S0RELH = 0x03;
                }
                    break;
        case 4800 :
                {
                    S0RELL = 0x98;
                    S0RELH = 0x03;
                }
                    break;
        case 2400 :
                {
                    S0RELL = 0x30;
```

```
                    S0RELH = 0x03;
                }
                break;
    default :
                {
                    S0RELL = 0xCC;
                    S0RELH = 0x03;
                }
                break;
    }
    return ;
}
/*************************************************************
/函数名称：timer0init()
/函数功能：初始化 Timer0
/输入参数：无
/返回参数：无
*************************************************************/
void timer0init(void)
{
    TMOD = 0x01;
    TH0 = (65536 - 50000)/256;
    TL0 = (65536 - 50000)%256;
    ET0 = 1;
    TR0 = 1;
}
/*************************************************************
/函数名称：workclkset()
/函数功能：nRF24LE1 工作时钟配置函数
/输入参数：无
/返回参数：无
*************************************************************/
void workclkset(void)
{
    CLKCTRL = 0x28;              /* 使用 XCOSC 16 MHz 的主时钟 */
    CLKLFCTRL = 0x01;
}
```

3. 功耗和时钟管理示例程序的串口功能函数

功耗和时钟管理示例程序的串口功能函数如下：

/***
/函数名称：putch()
/函数功能：串口发送一个字符
/输入参数：ch 为串口发送的字符
/返回参数：无
***/
void putch(char ch)
{
S0BUF = ch;
while(!TI0);
TI0 = 0;
}
/***
/函数名称：prints()
/函数功能：串口发送一个字符串
/输入参数：str 为指向发送的字符串的指针
/返回参数：无
***/
void prints(char * str)
{
while(* str != '\0')
  {
   putch(* str ++);
  }
}
/***
/函数名称：nextline()
/函数功能：串口发送换行显示信号
/输入参数：无
/返回参数：无
***/
void nextline(void)
{
 putch('\n');
}
/***
/函数名称：programdebug()
/函数功能：串口显示程序调试信息
/输入参数：bugs 为调试显示的信息的指针

```
/返回参数:无
***************************************************************/
void programdebug(char * bugs)
{
 nextline();
 prints("Debug Information:");
 prints((char *)bugs);
}
```

4. 功耗和时钟管理示例程序的低功耗模式相关功能函数

功耗和时钟管理示例程序的低功耗模式相关功能函数如下:

```
/***************************************************************
/函数名称:powrdownmodeset()
/函数功能:nRF24LE1 的低功耗模式设置
/输入参数:mode 为设置的低功耗模式
/返回参数:无
***************************************************************/
void powrdownmodeset(unsigned char mode)
{
  char low_power = 0;
  switch(mode)
  {
   case DeepSleep:low_power = 0x01;break;
   case Mem_Ret_On :low_power = 0x03;break;
   case Mem_Ret_Off:low_power = 0x02;break;
   case Reg_Ret_On :low_power = 0x04;break;
   case Reg_Ret_Off:low_power = 0x04;break;
   case Standby:low_power = 0x07;break;
   default:low_power = 0x07;break;
  }
  CLKCTRL = 0x10;               /*进入到睡眠前一定要启用 RC 时钟源*/
  OPMCON |= 0x02;
  PWRDWN |= low_power;
}
/***************************************************************
/函数名称:retentionpinset()
/函数功能:引脚唤醒模式设置
```

/输入参数:无
/返回参数:无
***/

```c
void (void)
{
  WUOPC0 = 0x00;
  OPMCON = 0x00;              /* 开锁 */
  WUOPC1 = 0x00;              /* 唤醒的引脚设置无 */
  WUOPC0 = 0x01;              /* 唤醒的引脚设置为 P0.0 */
  P0DIR |= 0x01;              /* P0.0 初始化为输入 I/O 口 */
  P00 = 0x00;                 /* P0.0 初始化为高电平 */
}
```

/**
/函数名称:getlastpowerdownmode()
/函数功能:获得上一次低功耗模式,并串口显示相关信息
/输入参数:无
/返回参数:无
***/

```c
void getlastpowerdownmode(void)
{
  char tmp = 0;
  tmp = PWRDWN&0x03;
  nextline();
  switch(tmp)
  {
    case 0x00:prints("Power Off!"); break;
    case 0x01:prints("Deep Sleep!");break;
    case 0x02:prints("Memory Retention,Timer Off!");break;
    case 0x03:prints("Memory Retention,Timer On!") ;break;
    case 0x04:prints("Register Retention!");break;
    case 0x07:prints("Standby!");break;
    default   :prints("Error Or Reserved!");break;
  }
}
```

/**
/函数名称:getlastresetinf()
/函数功能:获取并串口显示上一次复位的原因
/输入参数:无

```
/返回参数:无
*****************************************************/
void getlastresetinf(void)
{
 char tmp = 0;
 tmp = RSTREAS&0x07;                /*读取上一次复位状态保存寄存器*/
 nextline();
 prints("Reset Reason:");           /*显示复位原因*/
 if(0 == tmp)
    {
    prints("On - chip reset generator!");
    }
 else if(0x01 == tmp)
    {
    prints("RST pin!");
    }
 else if(0x02 == tmp)
    {
    prints("Watchdog!");
    }
 else if(0x04 == tmp)
    {
    prints("Reset from on - chip hardware debugger!");
    }
 else if(0x06 == tmp)
    {
    prints(" a reset from the debugger followed by a watchdog reset!");
    }
}
```

5. 功耗和时钟管理示例程序的中断服务函数

功耗和时钟管理示例程序的中断服务函数如下:

```
/*****************************************************
/函数名称:timer0isr()
/函数功能:处理Timer0中断
/输入参数:无
/返回参数:无
```

```c
                      ****************************************/
void timer0isr() interrupt TF0
{
  static char times = 0;
  TR0 = 0;
  TH0 = (65536 - 50000)/256;          /* 关闭定时器 */
  TL0 = (65536 - 50000) % 256;        /* 重新载入定时值 */
  times ++ ;
  if(20 == times)
   {
    LED = !LED;
    times = 0;
    seconds ++ ;
    nextline();
    programdebug("Time Left:");        /* 秒数累加 */
    putch(((5 - seconds) + '0'))  ;    /* 显示倒计时 */
    if(5 == seconds)
      {
       delay(100);
       programdebug("Sleeping..!");
       powrdownmodeset(DeepSleep);     /* 时间计为 0,进入睡眠 */
      }
    if(60 == seconds)
      {seconds = 0;}
   }
  TR0 = 1;                             /* 打开定时器 */
}
/**************************************************************
/函数名称:chipiocfg()
/函数功能:初始化 nRF24LE1 的 I/O 口
/输入参数:无
/返回参数:无
***************************************************************/
void chipiocfg(void)
{
  P0DIR &= 0x78;
  P07 = 0;
}
```

6. 功耗和时钟管理示例程序的主函数

功耗和时钟管理示例程序的主函数如下：

```
/*************************************************
/主函数部分
*************************************************/
void main(void)
{
  char flag = 1;
  Disableint();
  workclkset();                    /*初始化工作时钟*/
  baudrate(9600);                  /*波特率初始化为9 600*/
  chipiocfg();                     /*nRF24LE1的I/O口配置函数*/
  timer0init();                    /*定时器初始化函数*/
  delay(100);                      /*延时函数*/
  getlastresetinf();               /*获取上一次系统复位的原因*/
  nextline();
  getlastpowerdownmode();          /*获取上一次低功耗模式*/
  retentionpinset();               /*设置唤醒引脚*/
  delay(100);
  Enableint();
  while(1)
  {
    if(flag)
    {
      flag = 0;
      delay(100);
      getlastpowerdownmode();
    }
  }
}
```

2.9 电源监控

2.9.1 电源监控结构与功能

电源监控部分在上电时初始化系统，提供电源失效预警信号，并在电源电压过低时将系统复位,以确保系统的安全。

电源监控部分具有带延时的上电复位功能、在系统所有模式下的掉电复位功能、可编程阈值的电源失效报警,以及中断和硬件保护程序存储器中数据的功能。

电源监控部分方框图如图 2.9.1 所示。

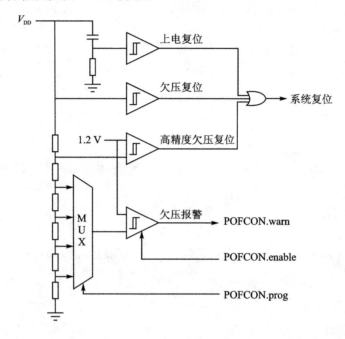

图 2.9.1　电源监控部分方框图

1. 上电复位

上电复位电路(POR)在上电时初始化系统。该电路由一个 RC 网络和一个比较器组成。为保证正常的工作,电源电压应调单上升,上升时间应满足系统要求。在供电电压达到 1.9 V 后,系统应保持复位状态至少 1 ms。上电复位时间如图 2.9.2 所示。

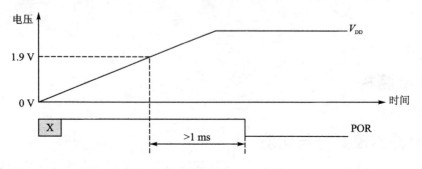

图 2.9.2　上电复位时间

2. 欠压复位

当电源电压跌落到欠压复位电路（BOR）所设定的阈值电压时，BOR将强制系统进入复位状态。BOR包含一个高精度比较器，该比较器在有效模式和待机模式下工作，阈值电压大约为1.8 V，大约有50 mV的滞后现象（V_{HYST}）。也就是说，当电源电压低于1.8 V时，触发系统复位时，电源电压必须恢复到1.85 V以上才可使nRF24LE1进入正常工作状态。滞后作用可防止当供电电压接近阈值电压时，比较器产生振荡现象。低功耗比较器的典型阈值电压是1.5 V。欠压复位示意图如图2.9.3所示。

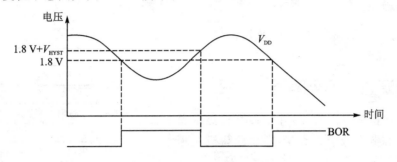

图 2.9.3　欠压复位示意图

3. 电源失效比较器

电源失效比较器（POF）可以给MCU提供一个电源故障的早期预警，它不会复位系统，但可以提供时间，使MCU可以进行有序的掉电操作，同时为程序存储器中的数据提供硬件保护，禁止写指令操作。

通过对POFCON寄存器的enable位的设置，可以实现POF比较器的使能或禁止。POF比较器使能时，当上电时，系统在有效或待机模式下工作。当电源电压低于可编程的阈值电压时，warn位被置1，同时也产生一个中断（POFIRQ）。warn位被置1时，程序存储器写指令将被禁止。电源失效比较器示意图如图2.9.4所示。

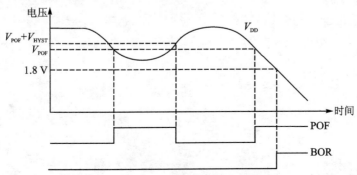

图 2.9.4　电源失效比较器示意图

第2章 nRF24LE1 的 MCU 与应用

可以使用 prog 位来配置所需的阈值电压(V_{POF}),可选的阈值电压包括 2.1 V、2.3 V、2.5 V 和 2.7 V,即电源电压下降的阈值。比较器将会有大约十几 mV 的滞后效应(V_{HYST})。

4. 电源监控特殊功能寄存器 POFCON

POFCON 寄存器完成 POF 的使能、阈值电压和 POF 报警功能的设置。有关 POFCON 寄存器的更多内容请登录 www.nordicsemi.com,查询 nRF24LE1 Ultra-low Power Wireless System On-Chip Solution Preliminary Product Specification v1.6。

2.9.2 电源监控示例程序流程图

本示例程序用来完成对电源的监控,程序流程图如图 2.9.5 所示。

2.9.3 电源监控示例程序

本示例程序检测电源电压,并在计算机显示器上显示。如果电源电压低于 2.1 V,则在计算机显示器上显示的电压低于 2.1 V;如果电源电压高于 2.1 V,则显示高于 2.1 V。示例程序的源代码如下:

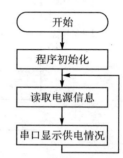

图 2.9.5 电源监控示例程序流程图

```
#include "reg24le1.h"
#define LED P00

/*******************************************************
/函数名称:delay()
/函数功能:软件延时
/输入参数:x 为延时的时间数
/返回参数:无
*******************************************************/
void delay(unsigned int x)
{
    unsigned int dy;
    for(;x>0;x--)
        for(dy=1000;dy>0;dy--)
        {
            ;
        }
}
/*******************************************************
/函数名称:IO_CONFIG()
```

```c
/函数功能:初始化 I/O 口
/输入参数:无
/返回参数:无
**********************************************************/
void IO_CONFIG()
{
    PODIR &= 0xFE;              /*配置P0.0为输出*/
    LED = 0;
}
/***********************************************************
/函数名称:PWR_CONFIG()
/函数功能:配置电压报警寄存器
/输入参数:无
/返回参数:无
**********************************************************/
void PWR_CONFIG()
{
    POFCON |= 0x80;             /*使能供电失败报警,设置比较电压为2.1 V*/
}
/***********************************************************
/函数名称:RD_POF()
/函数功能:读取电源失败的原因
/输入参数:无
/返回参数:无
**********************************************************/
unsigned char RD_POF()
{
    return (POFCON &= 0x10);    /*如果是0,则电压高于比较值;如果是1,则电压低于比较值*/
}
/***********************************************************
/函数名称:uart_init()
/函数功能:初始化串口
/输入参数:无
/返回参数:无
**********************************************************/
void uart_init()
{
    CLKCTRL = 0x28;             /*设置主时钟源为XCOSC 16 MHz*/
    CLKLFCTRL = 0x01;
    PODIR &= 0xF7;              /*配置P0.3(TXD)为输出*/
```

第 2 章　nRF24LE1 的 MCU 与应用

```
    PODIR |= 0x10;              /* 配置 P0.4(RXD)为输入 */
    P0 |= 0x18;
    S0CON = 0x50;
    PCON |= 0x80;               /* 设置波特率倍增 */
    WDCON |= 0x80;              /* 选用内部波特率发生器 */
    S0RELL = 0xFB;
    S0RELL = 0xF3;              /* 设置波特率为 38 400 */
}
/*******************************************************************
/函数名称：send()
/函数功能：通过串口发送一个字符到串口终端
/输入参数：ch 为串口发送的字符
/返回参数：无
*******************************************************************/
void send(char ch)              /* 通过串口发送一个字符 */
{
S0BUF = ch;
while(!TI0);                    /* 等待发送完成 */
TI0 = 0;
}
/*******************************************************************
/函数名称：puts()
/函数功能：发送一个字符串的函数
/输入参数：无
/返回参数：无
*******************************************************************/
void puts(char * str)
{
while( * str != '\0')
{
send( * str++ );
}
}
/*******************************************************************
/主函数部分
*******************************************************************/
void main()
{
uart_init();
PWR_CONFIG();
```

```
IO_CONFIG();
puts("power checking test!");      /*打印提示信息*/
send('\n');
while(1)
{
 if(RD_POF())                      /*检测反馈当前的供电电压情况*/
 puts("power votage is below 2.1");
 else
 puts("power votage is above 2.1");
 send('\n');
 delay(1000);
 }
}
```

程序运行后在计算机显示器上显示的结果如图 2.9.6 所示。本程序运行后,在计算机显示器上显示"power votage is above 2.1"。因为 nRF24LE1 正常供电是 3.3 V,3.3 V 高于 2.1 V,所以显示出如上的信息;否则会显示"power votage is below 2.1"。

图 2.9.6　程序运行后在计算机显示器上显示的结果

2.10 片上振荡器

nRF24LE1 包含两个高速振荡器和两个低速振荡器:一个是高速时钟源,即 16 MHz 晶体振荡器;另一个是用于快速启动的 16 MHz RC 振荡器,具有±5%频率精度。16 MHz RC 振荡器用来在等待主高速晶体振荡器时钟稳定前,提供系统高速时钟。低速时钟可以由 32.768 kHz 晶体振荡器或 32.768 kHz RC 振荡器提供,32.768 kHz RC 振荡器具有超低功耗和±10%的频率精度。使用外部的 16 MHz 和 32.768 kHz 时钟也可用来取代片内的时钟振荡器。

2.10.1 16 MHz 晶体振荡器

16 MHz 晶体振荡器(XOSC16M)电路如图 2.10.1 所示,采用 AT 切割的石英晶体振荡器,设计为并联谐振模式。为了得到准确的振荡频率,负载电容与晶体数据手册规格相匹配是很重要的。负载电容指的是从晶体来看外部所有等效电容之和。

$$C_{\text{LOAD}} = \frac{C'_1 \cdot C'_2}{C'_1 + C'_2}$$

$$C'_1 = C_1 + C_{\text{PCB1}} + C_{\text{PIN}}$$

$$C'_2 = C_2 + C_{\text{PCB2}} + C_{\text{PIN}}$$

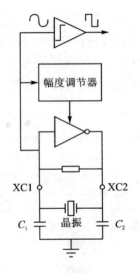

图 2.10.1 16 MHz 晶体振荡器电路

C_1 和 C_2 是陶瓷 SMD 电容,分别连在晶体引脚端和 V_{SS} 之间。C_{PCB1} 和 C_{PCB2} 是 PCB 上的杂散电容,而 C_{PIN} 是 nRF24LE1 的 XC1 和 XC2 引脚端的输入电容(典型值为 1 pF)。C_1 和 C_2 应取同一值(应尽可能接近)。

为了确保无线连接的功能,频率精度不能低于 $±60×10^{-6}$。应注意负载电容对晶体的初始精度、温漂、频率偏移的影响。当负载电容或并联电容较大时,推荐使用较低 ESR 的晶体,以提供更快的启动时间和较低的电流消耗。

当采用 12 pF 负载电容、3 pF 并联电容及 ESR 为 50 Ω 的晶体时,典型的启动时间为 1 ms。

晶体振荡器通常是在有效和待机模式下工作,通过写 1 到 CLKCTRL 寄存器的第 7 位,也可使其在寄存器维持模式下工作。推荐系统期望在 5 ms 时间内再次唤醒时使用。原因是振荡器多工作几 ms 所需的功耗要低于重新启动系统所需要的功耗。

2.10.2　16 MHz RC 振荡器

16 MHz RC 振荡器(RCOSC16M)在晶体振荡器启动时为系统提供高速时钟。16 MHz RC 振荡器可在几 ms 内启动,精度为±5%。

在默认状态下,16 MHz RC 和晶体振荡器同时启动,在晶体振荡器稳定工作前,由 RC 振荡器提供系统时钟,然后系统自动切换到由晶体振荡器提供时钟,并关闭 RC 振荡器来降低功耗。CLKCTRL 寄存器中的位[3]可用来查询当前所使用的是哪一个振荡器。

系统也可配置为启动时只使用一个时钟。通过置位 CLKCTRL 的位[4]和位[5]来选择使用哪一个振荡器。注意,射频收发部分不能使用 RC 振荡器作为时钟源,并且若 RC 振荡器作为时钟源,ADC 的性能也会降低。

2.10.3　外部 16 MHz 时钟

nRF24LE1 也可使用 16 MHz 的外部时钟,外部时钟连接到 XC1 引脚端上。如果外部时钟是轨-轨的数字信号,写 1 到 CLKCTRL 寄存器的位[6]。输入的时钟信号也可以是模拟信号,例如来自微控制器的时钟振荡器信号,在这种情况下,nRF24LE1 的晶体振荡器必须使能,以便将输入的模拟信号转换为数字时钟信号。使能晶体振荡器,CLKCTRL 的位[6]必须写 0,CLKCTRL 的位[5:4]必须写 10。推荐的输入信号幅值为 0.8 V 或更高,以获得较好的低功耗性能和信噪比。输入信号的直流电平并不重要,只要信号电平不高于 V_{DD} 或低于 V_{SS} 即可。XC1 引脚端将会给微控制器晶体增加约 1 pF 的负载电容,以及 PCB 走线的附加杂散电容。XC2 引脚端不需要连接。

注意:为了获得可靠的无线连接,频率精度应不低于 $±60×10^{-6}$。

2.10.4　32.768 kHz 晶体振荡器

除了深度睡眠模式和存储器维持定时器关闭模式外,32.768 kHz 晶体振荡器(XOSC32K)可在其他所有模式下工作,通过写 000 到 CLKLFCTRL 的位[2:0]来使能。32.768 kHz 晶体振荡器电路如图 2.10.2 所示。

晶体必须连接在 P0.0 和 P0.1 之间,当振荡器使能后,该引脚端自动配置为振荡器引脚。与 16 MHz 晶体振荡器(XOSC16M)类似,为了得到准确的振荡频率,负载电容与晶体数据手册规格相匹配是很重要的。

C_1 和 C_2 是陶瓷 SMD 电容,分别连在晶体引脚端和 V_{SS} 之间。C_{PCB1} 和 C_{PCB2} 是 PCB 上的杂散电容,而 C_{PIN} 是

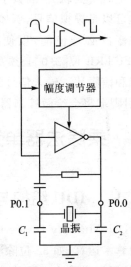

图 2.10.2　32.768 kHz 晶体振荡器电路

nRF24LE1 的 P0.0 和 P0.1 引脚端的输入电容（典型值为 3 pF）。C_1 和 C_2 应取同一值（应尽可能接近）。

当采用 9 pF 负载电容、3 pF 并联电容及 ESR 为 50 Ω 的晶体时，典型的启动时间为 0.5 s。

查询 CLKLFCTRL 寄存器的位[6]可以得知振荡器是否已经就绪。

2.10.5　32.768 kHz RC 振荡器

低速时钟也可由 32.768 kHz RC 振荡器（RCOSC32K）替代晶体振荡器，其精度是 ±10%，此时 P0.0 和 P0.1 可作其他应用。32.768 kHz RC 振荡器通过写 001 到 CLKLFCTRL 的位[2:0]来使能。典型的启动时间为 0.5 ms。查询 CLKLFCTRL 寄存器的位[6]可以得知振荡器是否已经就绪。

2.10.6　合成 32.768 kHz 时钟

低速时钟也可由 16 MHz 晶体振荡器时钟来合成，即写 010 到 CLKLFCTRL 的位[2:0]来选择此功能。合成时钟只能在 16 MHz 晶体振荡器有效的系统模式下使用（有效工作模式包括有效模式、待机模式和寄存器维持模式）。

2.10.7　外部 32.768 kHz 时钟

nRF24LE1 可以使用 32.768 kHz 外部时钟，32.768 kHz 外部时钟连接到 P0.1 引脚端。如果外部时钟是轨-轨的数字信号，则写 100 到 CLKLFCTRL 寄存器的位[2:0]。输入的时钟信号也可以是模拟信号，例如来自微控制器的时钟振荡器信号，在这种情况下，nRF24LE1 的晶体振荡器必须使能，以便将输入的模拟信号转换为数字时钟信号。使能晶体振荡器，CLKLFCTRL 位[2:0]必须写 011。推荐的输入信号幅值为 0.2 V 或更高，以获得较好的低功耗性能和信噪比。输入信号的直流电平并不重要，只要信号电平不高于 V_{DD} 或低于 V_{SS} 即可。P0.1 引脚端将会给微控制器晶体增加约 3 pF 的负载电容，以及 PCB 走线的附加杂散电容。

2.11　乘除法器单元 MDU

2.11.1　MDU 结构与功能

MDU（Multiply Divide Unit，乘除法单元）是一个片内的算术协处理器，使 MCU 可完成扩展的算术运算，如 32 位除法、16 位乘法、移位和标准操作。

MDU 功能由 MDU 特殊功能寄存器 MD0～MD5 和 ARCON 控制。与 MDU 有关的寄存器如图 2.11.1 所示。

所有操作都是无符号整数操作,MDU 由 7 个特殊功能寄存器来处理。算术单元允许独立工作,可与 MCU 部分并行工作。

操作数和结果存储在 MD0～MD5 寄存器中,整个单元由 ARCON 控制,任何 MDU 运算将覆盖其操作数。

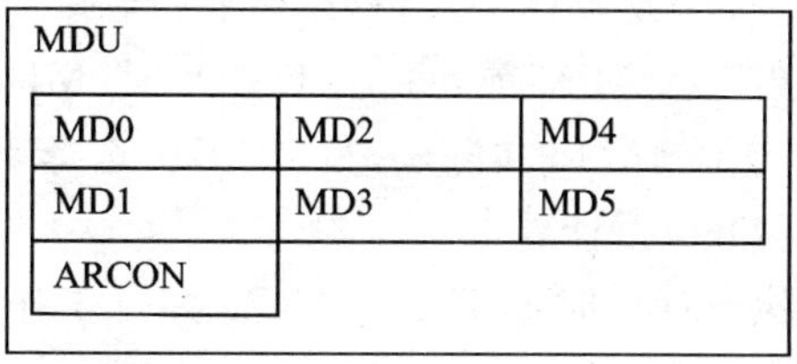

MDU 不允许重入代码,不能被主程序和中断程序同时使用。使用 NOMDU_R515 编译指令,可以禁止 MDU 操作,以避免可能的函数冲突。

MDU		
MD0	MD2	MD4
MD1	MD3	MD5
ARCON		

图 2.11.1　与 MDU 有关的寄存器

2.11.2　MDU 操作步骤

1. 寄存器 MD0～MD5

寄存器 MD0～MD5 在 MDU 工作时使用。其地址分配为 0xE9～0xEE。操作数和结果存储在 MD0～MD5 寄存器中。

2. ARCON 寄存器

ARCON 寄存器控制 MDU 的操作并指示其当前状态。

3. MDU 的操作步骤

(1) 装载 MDx 寄存器

MDU 执行的计算类型由 MDx 寄存器写入的顺序来选择。MDU 寄存器写入顺序如表 2.11.1 所列。

表 2.11.1　MDU 寄存器写入顺序

选择	32/16 位		16/16 位		16 位×16 位		移位/规范化	
最先写	MD0 (LSB) MD1 MD2 MD3 (MSB)	被除数	MD0 (LSB) MD1 (MSB)	被除数	MD0 (LSB) MD4 (LSB)	数 1 数 2	MD0 (LSB) MD1 MD2 MD3 (MSB)	数
最后写	MD4 (LSB) MD5 (MSB)	除数	MD4 (LSB) MD5 (MSB)	除数	MD1 (MSB) MD5(MSB)	数 3 数 4	ARCON	

① 写 MD0 启动任一种运算;
② 写操作,由表 2.11.1 来确定 MDU 的运算;
③ 写入 MD5 或 ARCUN 启动选中的运算。

SFR 控制检测到上述顺序并将控制转移给 MDU。当在写访问到 MD0～MD5 之间时,写

第 2 章 nRF24LE1 的 MCU 与应用

访问 MD2 或 MD3,选择 32/16 位除法。

当写入 MD5 前,写访问 MD4 或 MD1,选择 16/16 位除法或 16 位×16 位乘法。写入 MD4 选择 16/16 位除法,写入 MD1 选择 16 位×16 位乘法,即 Num1(数 1)×Num2(数 2)。

(2) 执行计算

执行计算时,MDU 与 MCU 并行工作。MDU 的不同运算操作执行时间不同,操作执行时间为 3~19 个时钟周期。

(3) 从 MDx 寄存器读出结果

MDU 寄存器读出顺序如表 2.11.2 所列。

表 2.11.2 MDU 寄存器读出顺序

选 择	32/16 位		16/16 位		16 位×16 位		移位/标准	
最先读	MD0 (LSB) MD1 MD2 MD3 (MSB)	商	MD0 (LSB) MD1 (MSB)	商	MD0 (LSB) MD4 (LSB)	乘积	MD0 (LSB) MD1 MD2	数
最后读	MD4 (LSB) MD5 (MSB)	余数	MD4 (LSB) MD5 (MSB)	余数	MD1 (MSB) MD5 (MSB)		MD3 (MSB)	

读出的第一个 MDx 寄存器的顺序不是关键,但最后(从 MD5-除法、MD3-乘法、移位或规范化)的读确定整个计算的结束(第(3)步结束)。

(4) 规范化

所有存储在 MD0~MD3 的 32 位整型变量前面的 0 都将通过左移操作删除,当 MD3 寄存器的最高位为 1 时整个操作完成。规范化以后,位 ARCON[4](MSB)~ARCON[0](LSB) 包含已经完成的左移操作的数。

(5) 移 位

在执行移位操作时,存储在 MD0~MD3 的 32 位整型变量(后面包含最高位字节)将左移或右移指定的位数。方向位(ARCON[5])定义移位的方向,而 ARCON[4:0] 指定移位的位数(必须非 0)。在移位操作期间,右移时 0 将进入 MD3 的最左端,左移时将进入 MD0 的最右端。

(6) mdef 标志

mdef 错误标志指示产生的一个不正确执行的操作(当一个算术运算重新启动或被一个新的操作所中断时)。当第一次写操作到 MD0 时,错误标志机制自动使能,而当在第(3)步最后读出 MD3(乘法或移位/规范化)或 MD5(除法)后禁止。

当以下情况出现时,错误标志置位:

- 在执行第(2)步 MDU 操作时,写 MD0~MD5 或 ARCON(重新启动或计算中断);
- 在执行第(2)步 MDU 操作时,若错误标志机制使能,则任何读出 MDx 寄存器的操作都将导致错误标志置位。在此情况下,错误标志置位,但计算不会中断。
当读访问 ARCON 寄存器后,错误标志复位。错误标志位只读。

(7) mdov 溢出标志

mdov 溢出标志在以下情况发生时置位:
- 被 0 除;
- 乘法结果大于 0000 FFFFH;
- 启动规范化操作时,如果 MD3 的最高位为 1(MD3[7]=1)。

MDU 任何不符合上述条件的操作将清除溢出标志。

注意:溢出标志由专门硬件控制,不可写。

2.11.3　MDU 示例程序流程图

nRF24LE1 的 MDU 可以实现 16 位的乘法、除法、32 位的除法和移位运算等。在本示例程序里利用了 MDU 来实现 16 位的除法。通过把除数和被除数按照一定的先后顺序送入到 MDU,然后再从 MDU 读出最后的商和余数。

本示例程序的流程图如图 2.11.2 所示。

2.11.4　MDU 示例程序

1. MDU 示例程序的宏定义

MDU 示例程序的宏定义如下:

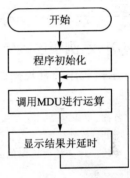

图 2.11.2　MDU 示例程序流程图

```
#include "reg24le1.h"
/****************************************************
/用到的一些宏定义
*****************************************************/
#define toasc(x)      (x+'0')       /*数字转换成对应的 ASCII 码*/
#define UPtoLOW(x)    (x-'A'+'a')   /*大写转换成小写*/
#define LOWtoUP(x)    (x-'a'+'A')   /*小写转换成大写*/
#define LED P10                      /*定义一个工作状态指示灯(LED)*/
#define LEFT   0x10                  /*定义一个方向,左*/
#define RIGHT  0x01                  /*定义一个方向,右*/
#define true   0x01
#define false  0x00
/****************************************************
/函数原型声明开始
```

```
*******************************************************************/
void ioconfig();
void delay(unsigned int dj);
void uart();
void putch(char ch);
void puts(char * str);
unsigned int MDUmul_16bit(unsigned int x1,unsigned int x2);
void MDUdiv_16bit(unsigned int x1,unsigned int x2,unsigned int * Quotient,unsigned int
* Remainder );
void MDUdiv_32bit(unsigned int x1,unsigned int x2,unsigned int * Quotient,unsigned int
* Remainder );
unsigned int shiffunc(unsigned int dat,unsigned char bitnum,unsigned char derec);
void display(unsigned int x);
/************************函数原型声明结束************************/
```

2. MDU 示例程序的常规操作函数

MDU 示例程序的常规操作函数如下：

```
/*******************************************************************
/函数名称：delay()
/函数功能：软件延时
/输入参数：dj 为软件延时的时间数
/返回参数：无
*******************************************************************/
void delay(unsigned int dj)
{
unsigned char di;
 for(;dj>0;dj--)
   for(di=120;di>0;di--)
   {
     ;
   }
}
/*******************************************************************
/函数名称：ioconfig()
/函数功能：配置与 LED 灯连接的 I/O 为输出,初始电平为低电平
/输入参数：无
/返回参数：无
*******************************************************************/
void ioconfig()
```

```c
{
    P1DIR &= 0xFE;                    /* 配置 GPIO 为输出 */
    P10 = 0;
}
```

3. MDU 示例程序的串口功能函数

MDU 示例程序的串口功能函数如下:

```c
/***************************************************************
/函数名称: uart()
/函数功能: 初始化串口
/输入参数: 无
/返回参数: 无
***************************************************************/
void uart()
{
    CLKCTRL = 0x28;                   /* MCU 时钟源设置 */
    CLKLFCTRL = 0x01;                 /* 设置 32.768 kHz 实时时钟 */
    P0DIR &= 0xF7;                    /* 配置 P0.3(TXD)为输出 */
    P0DIR |= 0x10;                    /* 配置 P0.4(RXD)为输入 */
    P0 |= 0x18;
    S0CON = 0x50;
    PCON |= 0x80;                     /* 波特率倍增 */
    WDCON |= 0x80;                    /* 选择内部波特率发生器 */
    S0RELL = 0xFB;
    S0RELL = 0xF3;                    /* 波特率设置为 38 400 */
}
/***************************************************************
/函数名称: putch()
/函数功能: 串口发送一个字符
/输入参数: ch 为发送的字符
/返回参数: 无
***************************************************************/
void putch(char ch)
{
    S0BUF = ch;
    while(!TI0);
    TI0 = 0;
}
```

```
/*******************************************************
/函数名称:puts()
/函数功能:串口发送一个字符串
/输入参数:str 为指向字符串的指针
/返回参数:无
*******************************************************/
void puts(char * str)
{
while( * str != '\0')
{
putch( * str ++ );
}
}
```

4. MDU 示例程序的协处理单元功能函数

MDU 示例程序的协处理单元功能函数如下:

```
/*******************************************************
/函数名称:MDUmul_16bit()
/函数功能:MDU 的 16 位乘法
/输入参数:乘数 x1,乘数 x2
/返回参数:乘积
*******************************************************/
unsigned int MDUmul_16bit(unsigned int x1,unsigned int x2)
{
unsigned int res = 0;
MD0 = x1&0xff;                    /* 写入低 8 位数据 */
MD4 = x2&0xff;
MD1 = ((x1&0xff00) >> 8);
MD5 = ((x2&0xff00) >> 8);
delay(10);
res += MD0;
res += MD3 * 256;
return res;
}
/*******************************************************
/函数名称:MDUdiv_16bit()
/函数功能:16 位数的除法
/输入参数:x1 为被除数,x2 为除数,Quotient 为计算的商,Remainder 为计算余数
```

/返回参数:无
**/

```c
void MDUdiv_16bit(unsigned int x1,unsigned int x2,unsigned int * Quotient,unsigned int
    * Remainder )
{
    * Quotient = 0;
    * Remainder = 0;
    MD0 = x1&0xff;                      /*写入低8位*/
    MD1 = ((x1&0xff00) >> 8);
    MD4 = x2&0xff;
    MD5 = ((x2&0xff00) >> 8);
    delay(10);
    * Quotient = MD0 + MD1 * 256;       /*读取除法的商*/
    * Remainder += MD4;
    * Remainder += MD5 * 256;           /*读取除法的余数*/
}
```

/**
/函数名称:MDUdiv_32bit()
/函数功能:32位数据的除法函数
/输入参数:x1为被除数,x2为除数,Quotient为计算的商,Remainder为余数
/返回参数:无
**/

```c
void MDUdiv_32bit(unsigned int x1,unsigned int x2,unsigned int * Quotient,unsigned int
    * Remainder )
{
    * Quotient = 0;
    * Remainder = 0;
    MD0 = x1&0xff;                      /*写入除数的低8位*/
    MD1 = (x1 >> 8)&0xff;               /*写入除数的高8位*/
    MD2 = (x1 >> 16)&0xff;
    MD3 = (x1 >> 24)&0xff;              /*写入最高的8位*/
    MD4 = x2&0xff;
    MD5 = (x2 >> 8)&0xff;
    delay(10);
    * Quotient = MD0 + MD3 * 256;       /*读取除法的商*/
    * Remainder += MD4;
    * Remainder += MD5 * 256;           /*读取除法的余数*/
}
```

```
/******************************************************************
/函数名称：shiffunc()
/函数功能：移位功能函数
/输入参数：dat 为需要移位的数据，bitnum 为移位的位数，derec 为移位的方向
/返回参数：移位后的结果
******************************************************************/
unsigned int shiffunc(unsigned int dat,unsigned char bitnum,unsigned char derec)
{
  unsigned int res = 0;
  MD0 = dat&0xff;                /*写入要移位的数据的低8位*/
  MD3 = (dat >> 8)&0xff;         /*写入要移位的数据的高8位*/
  if(derec == LEFT)
  ARCON &= ~(0x20);              /*左移*/
  else
  ARCON |= (0x20);               /*右移*/
  ARCON |= bitnum;               /*位[4:0]决定移位的位数*/
  delay(5);
  res += MD0;
  res += MD3 * 256;
  return res;                    /*返回移位后得到的结果*/
}
```

5. MDU 示例程序的显示函数

MDU 示例程序的显示函数如下：

```
/******************************************************************
/函数名称：display()
/函数功能：显示函数将数字分解成4位的ASCII码发送给计算机
/输入参数：x 为串口显示的字符
/返回参数：无
******************************************************************/
void display(unsigned int x)
{
unsigned char buf[8];            /*存储分离的位的数组定义*/
unsigned char bitw = 0;
unsigned char tmph = 0,tmpl = 0;
while(x)
{
  buf[bitw++] = x&0xff;          /*读取数据低8位*/
  x >>= 8;
```

```
    }
    if(bitw == 0)
    {
    putch('\n');
    puts(" the res == ");              /*串口打印数据提示字符串*/
    putch(toasc(0));
    }
    else
    {
    while(bitw)                        /*计算有效数据位宽*/
    {
    putch('\n');
    puts(" the res == 0x");            /*显示数据输出的字符串提示*/
    if((buf[bitw-1]/16)>9)             /*十进制数据无法显示*/
    {tmph = (buf[bitw-1]/16) - 10 + 'a'; }   /*转换成十六进制字母显示*/
    else
    {putch(toasc(buf[bitw-1]/16)); }
    if((buf[bitw-1]%16)>9)
    {tmph = (buf[bitw-1]%16) - 10 + 'a'; }
    else
    {putch(toasc(buf[bitw-1]%16)); }
    bitw--;
    }
    }
    putch('\n');                        /*换行*/
    }
```

6. MDU 示例程序的主函数

MDU 示例程序的主函数如下:

```
/*************************************************************
/主函数部分
**************************************************************/

void main()
{
unsigned int datq,datr;
uart();
ioconfig();
delay(10);
```

```c
    puts("this is just a test of MDU....");   /*显示提示信息*/
    putch('\n');
    while(1)
    {
      MDUdiv_16bit(1024,512,&datq,&datr);    /*验证MDU的16位除法单元功能
                                                1 024/512,datq保存商,datr保存余数*/

      putch(datq+'0');                       /*串口打印商,datq为1 024/512=2*/
      putch(' ');
      putch(datr+'0');                       /*串口打印余数,余数为0*/
      LED = !LED;                            /*LED闪烁*/
      putch('\n');
      delay(10000);                          /*延时*/
    }
}
```

程序运行结果:

在主函数部分,只调用了1个做16位除法的函数,函数中设置的被除数是1 024,除数是512,运算得到的商保存在datq中,余数保存在datr中。运算的结果(商为2,余数为0)通过串口显示在计算机的显示器上,如图2.11.3所示。

图2.11.3 串口数据截图

2.12 加密/解密协处理器

使用片内的固件加密/解密协处理器可以大大提高软件的时间和能耗效率。固件加密/解密协处理器是一个 8 位输出的 Galois Field 乘法器,使用下列多项式:

$$m(x) = x^8 + x^4 + x^3 + x + 1$$

这是 AES(Advanced Encryption Standard,高级加密标准)使用的多项式。

加密/解密协处理器的固件程序可由 Nordic 所提供,当输入数据寄存器发生改变时,在一个时钟周期内即可获得结果。

加密/解密协处理器结构方框图如图 2.12.1 所示。

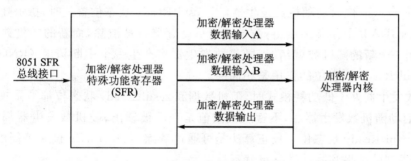

图 2.12.1 加密/解密协处理器结构方框图

CCPDATIA、CCPDATIB 和 CCPDATO 寄存器控制加密/解密协处理器,地址为 0xDD~0xDF。寄存器 CCPDATIA 和 CCPDATIB 为数据输入寄存器(读/写)。寄存器 CCPDATO 为数据输出寄存器(只读),数据为来自协处理器的输出结果。当输入数据寄存器发生改变时,CCPDATO 寄存器将在一个时钟周期内更新结果。

有关加密/解密协处理器的更多内容请登录 www.nordicsemi.com,查询 nRF24LE1 Ultra-low Power Wireless System On-Chip Solution Preliminary Product Specification v1.6。

2.13 随机数发生器

2.13.1 随机数发生器结构与功能

nRF24LE1 包含一个基于热噪声原理的非确定结构算法的随机数发生器(RNG),不需要种子值,采用热噪声来产生一个非确定的比特(数据)流和一个数字校正算法来移除比特流中趋向于 1 或 0 的数据,校正算法确保统计学的均匀分布,经过校正的数据顺序进入并行读出的 8 位寄存器。最高速率为 10 KBps。微控制器待机时,随机数发生器仍可工作。

第 2 章 nRF24LE1 的 MCU 与应用

随机数发生器结构方框图如图 2.13.1 所示。

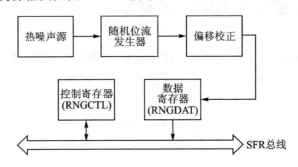

图 2.13.1 随机数发生器结构方框图

写 1 到 powerUp 位启动随机数发生器,当 resultReady 状态位置位时,指示随机数已经就绪,可以在 RNGDAT 寄存器中读出,数据读出后寄存器将被清除,而新的随机数据就绪时再次置位。每当一个新的随机数就绪时,随机数发生器将产生一个中断请求（RNGIRQ）,中断的作用与 resultReady 状态位的作用是相同的。

当随机数发生器处于低功耗模式时,随机数据和 resultReady 状态位都是无效的。写 1 到 powerUp 位启动随机数发生器后,不管随机数值是否已被读出,随机数发生器都将首先清除随机数据和 resultReady 状态位。校正算法是可通过清除 correctorEn 位来关闭的,这样可以提高速度,但得到的结果其统计学分布就不是很完美了。

产生一字节数据所需的时间是不可预知的,而且一字节到下一字节的时间也会发生变化,使能校正算法时尤其如此。当校正算法禁止时,平均产生一字节随机数需要 0.1 ms,使能校正算法时所需时间约为其 4 倍。在 powerUp 位置位并启动随机数发生器后,产生第一字节所需的时间约为 0.25 ms,随机数发生器通过两个寄存器 RNGCTL 和 RNGDAT 来控制。RNGCTL 包含控制位和状态位,RNGDAT 包含产生的随机数据。

2.13.2 随机数发生器示例程序流程图

本示例程序利用 nRF24LE1 内部的随机数发生器来产生随机数,并且通过串口发送到计算机显示器上。串口波特率配置成 38 400,数据格式为 8 位数据位,1 位停止位,没有验证位。

本示例程序的程序流程图如图 2.13.2 所示。

2.13.3 随机数发生器示例程序

随机数发生器示例程序的源代码如下:

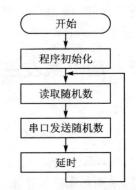

图 2.13.2 随机数发生器示例程序流程图

```c
#include "reg24le1.h"
#define RDMRD        (RNGCTL&0x20)          /*宏定义随机数发生器*/
/*************************************************************
/函数声明部分
*************************************************************/
void puts(char * str);
void delay(unsigned int dx);
void send(char ch);
void uart_init();
unsigned char readrdm(void);
void radmcof();
/*************************************************************
/函数名称: radmcof()
/函数功能: 初始化随机数发生器寄存器
/输入参数: 无
/返回参数: 无
*************************************************************/
void radmcof()
{
RNGCTL |= (0x80|0x40);                      /*配置随机数发生器*/
}
/*************************************************************
/函数名称: readrdm()
/函数功能: 读取生成的随机数的函数
/输入参数: 无
/返回参数: 生成的随机数
*************************************************************/
unsigned char readrdm(void)
{
while(!RDMRD);                              /*等待随机数的生成*/
return RNGDAT;                              /*返回生成好的随机数*/
}
/*************************************************************
/函数名称: init_uart()
/函数功能: 初始化 nRF24LE1 的串口
/输入参数: 无
/返回参数: 无
*************************************************************/
void init_uart(void)
{
```

```
        CLKCTRL = 0x28;                    /* 使用 XCOSC16M */
        CLKLFCTRL = 0x01;
        P0DIR &= 0xF7;                     /* 配置 P0.3(TXD)为输出 */
        P0DIR |= 0x10;                     /* 配置 P0.4(RXD)为输入 */
        P0 |= 0x18;
        S0CON = 0x50;
        PCON |= 0x80;                      /* 波特率倍增 */
        WDCON |= 0x80;                     /* 选用内部波特率发生器 */
        S0RELL = 0xFB;
        S0RELL = 0xF3;                     /* 设置波特率为 38 400 */
    }
```

/***
/函数名称：send()
/函数功能：通过串口发送一个字符的函数
/输入参数：ch 为串口发送的字符
/返回参数：无
***/

```
void send(char ch)
{
S0BUF = ch;                                /* 将要发送的数据写到串口的缓冲区 */
while(!TI0);                               /* 等待发送完成 */
TI0 = 0;
}
```

/***
/函数名称：delay()
/函数功能：软件延时
/输入参数：dx 为软件延时的时间数
/返回参数：无
***/

```
void delay(unsigned int dx)                /* 调用时通过设置参数 dx 来改变延时时间 */
{
unsigned int di;
    for(;dx>0;dx--)
      for(di = 120;di>0;di--)
          {
              ;
          }
}
```
/***

```c
/函数名称：puts()
/函数功能：串口显示一个字符串
/输入参数：str 为串口发送的字符串
/返回参数：无
******************************************************/
void puts(char * str)
{
    while( * str != '\0')                    /*如果没到字符串的末尾*/
    {
        send( * str ++ );                    /*通过连续调用 send()函数来实现字符串的打印*/
    }
}
/*****************************************************
/主函数部分
******************************************************/
void main()
{
    static unsigned char random;
    init_uart();                             /*串口的初始化*/
    radmcof();                               /*随机数发生器初始化*/
    delay(100);                              /*延时 100 时间单位*/
    send('\n');
    puts("...a test for Random number generator...");  /*输出程序启动的提示符*/
    send('\n');                              /*换行操作*/
    while(1)
    {
        random = readrdm();                  /*读取随机数*/
        send(' ');
        send((random/100) + '0');            /*显示随机数的百位*/
        send((random % 100)/10 + '0');       /*显示随机数的十位*/
        send((random % 10) + '0');           /*显示随机数的个位*/
        send(' ');                           /*显示空格*/
        delay(10000);                        /*延时*/
    }
}

/*****************************************************
/                    程序代码到此结束                    /
******************************************************/
```

第2章 nRF24LE1 的 MCU 与应用

程序运行结果:

在编译环境中将程序编译生成 HEX 文件,并通过下载器下载到 nRF24LE1 后,连上串口线,复位启动,在计算机显示器上,可以看见 nRF24LE1 产生的随机数(中间隔两个空格就会显示一个随机数),显示结果如图 2.13.3 所示。

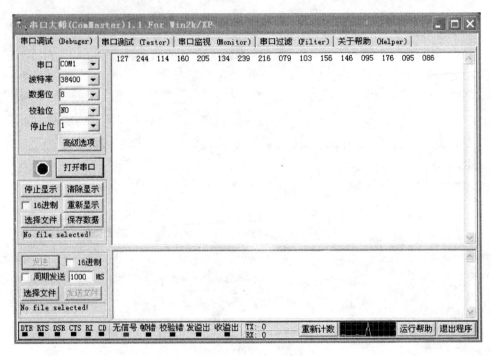

图 2.13.3　随机数串口显示截图

第 3 章

nRF24LE1 的接口与应用

3.1 通用 I/O 端口 GPIO

3.1.1 GPIO 结构与功能

1. 不同封装的 I/O 引脚

nRF24LE1 的 I/O 引脚端默认设置为 MCU GPIO(通用 I/O 端口)，不同的封装可用的 I/O 数目不同：24 引脚 4 mm×4 mm 封装是 7 个，32 引脚 5 mm×5 mm 封装是 15 个，48 引脚 7 mm×7 mm 封装是 31 个。这些 I/O 引脚端与部分外设模块功能是复用的，如 SPI、2 线接口、32 kHz 晶体振荡器和硬件调试的 JTAG 接口等。外设模块和引脚端的连接可用通过端口交叉模块(PortCrossbar)进行动态调整。每种封装的引脚分配如表 1.2.2、表 1.2.4 和表 1.2.5 所列。

nRF24LE1 的每个 I/O 引脚具有 MCU 通用的 I/O 引脚功能，包括数字或模拟、配置方向、配置驱动能力、配置上拉或下拉。这些功能是多路复用的，由端口交叉模块根据外设模块连接的需要来控制和配置。

nRF24LE1 的引脚连接到默认的引脚多路复用器(MUX)所连接的 MCU 通用 I/O 寄存器。寄存器 Pn.m (n 是端口数，m 是位数)包含 MCU 的通用 I/O 数据，寄存器 PDIRn.m 作输入和输出方向控制，寄存器 PCONn.m 控制每个引脚的驱动能力以及上拉/下拉电阻。

当 MCU 使能 nRF24LE1 的外设模块时，MUX 断开 MCU 对引脚的控制，并移交给端口交叉模块来设置引脚的方向和功能。然而，如果引脚作为模拟输入工作时，MCU 必须分别设置引脚控制寄存器 PDIR 和 PCON，以防止引脚分配的冲突，以及满足 nRF24LE1 模拟外设模块的需要。

2. Pn.m、PnDIRm 和 PnCONn 寄存器

nRF24LE1 的每个端口各有一个 Pn.m、PnDIRm 和 PnCONn。Pn.m 和 PnDIRm 各控制一个参数，这就意味着对其读/写操作将可分别直接控制读出端口的状态。然而，用 PnCONm

一次只能控制或读出一个引脚的功能。PnCON 包含了引脚的地址和输入/输出功能的信息,这些功能在使能时更新并启用。

PnCON 寄存器可提供以下功能:
- 输出缓冲器开启,提供正常驱动能力;
- 输出缓冲器开启,提供高驱动能力;
- 输入缓冲器开启,无上拉/下拉电阻;
- 输入缓冲器开启,有上拉电阻;
- 输入缓冲器开启,有下拉电阻;
- 输入缓冲器关闭。

例如:如果将端口 3 的 4 个引脚设置为带内部上拉电阻的输入引脚,则通过一个写操作到 P3DIR 和 4 个写操作到 P3CON,,每次写操作到 P3CON 仅更新引脚地址。

3. 端口交叉模块功能

端口交叉模块 I/O 引脚与外设模块间建立联系。

(1) 动态分配引脚

端口交叉模块可以根据系统需要实时修改与芯片的外设模块(如 SPI、2 线接口等)的动态连接。由于有效的引脚数量小于所有外设所需 I/O 总和,因此这个功能是必需的。在选用最小的封装时,在引脚的分配上可能会发生冲突。这可以通过设置每个外设模块的优先级来解决。

(2) 数字模块的动态引脚分配

每个数字外设模块接口需要对应 I/O 引脚,在分配表中指定了每个引脚,也同时列出了可能发生冲突的模块的优先级。如果模块被使能,并且没有优先级更高的模块被使能,则所有与该模块相关的 I/O 都将被该模块使用。端口交叉模块从不允许数字模块只使用部分数字 I/O 的请求,即使冲突仅在某些引脚存在。

(3) 模拟模块的动态引脚分配

模拟 I/O 的动态请求与数字 I/O 是类似的。由 ADCCON1 的 chsel 位和 ADCCON1 的 refsel 位配置,连接模拟信号到一个对应的引脚,输入到模拟模块。它与数字外设模块有所不同,数字外设模块要求一旦使能,则与模块相关的所有 I/O 都必须保留。两个模拟模块 ADC 和模拟比较器,在引脚分配表中共用一列,这是因为比较器使用 ADC 配置寄存器来为其信号和电压基准选择源引脚。

注意: 器件不能够预防数字外设模块和模拟外设模块同时使用同一引脚。如果一个引脚用作模拟输入,则应禁止数字 I/O 和数字外设模块连接到同一引脚。模拟模块间的冲突可以通过优先级来避免。

XOSC32K 需要的 I/O 是实时可编程的,取决于配置。这个模块可以要求需要模拟的或者数字的 I/O。如果一个引脚的模拟功能被使能,则没有修改或禁止引脚的数字配置。在使用模拟引脚时,如果部分数字输入/输出引脚是必需的,则在模拟模块使能前,必须通过 PxCON 和 PxDIR 单独使能这个配置。

(4) 默认引脚分配

如果没有外设模块请求需要 I/O,则在引脚分配表默认列的默认引脚被使能。这就意味着所有器件引脚将作为 MCU 的通用 I/O 使用。复位后,所有 I/O 配置为数字输入。引脚的功能、方向和 I/O 引脚上的数据由寄存器 PnCON、PnDIR 和 Pn 来控制。

默认引脚也包括条件使能的连接,即基于引脚方向的设置。例如,在 24 引脚 4 mm × 4 mm 封装芯片中,如果 P0DIR 寄存器设置 P0.6 为输入,则可作为 MCU 的通用输入和作为 UART 的接收;如果 P0.5 编程为输出,则可连接到 MCU 的通用输出,但同时也可通过一个"与"门作为有条件的 UART/TXD 输出。

3.1.2 I/O 端口可编程寄存器

nRF24LE1 可用的 4 个端口取决于封装类型。为使引脚实现期望的方向和功能,需要使用以下配置寄存器进行配置：PxDIR(P0DIR、P1DIR、P2DIR 和 P3DIR)、PxCON(P0CON、P1CON、P2CON 和 P3CON)。PxDIR 寄存器确定引脚的方向,PxCON 寄存器包含输入/输出操作时的功能选项。端口交叉模块复位后,默认所有引脚为输入,并连接到 MCU 的通用 I/O (pxDi)。

1. PxDIR 寄存器

为改变引脚方向,可将期望的方向值写入 PxDIR 寄存器。PxDIR 寄存器功能如表 3.1.1～表 3.1.4 所列。

表 3.1.1　P0DIR 寄存器(地址—0x93；复位值—0xFF)

位	名称	R/W	功能
[7:0]	dir	R/W	确定引脚端 P0.0～P0.7 方向。输出 dir=0,输入 dir=1 P0DIR0—P0.0 P0DIR1—P0.1 P0DIR2—P0.2 P0DIR3—P0.3 P0DIR4—P0.4 P0DIR5—P0.5 P0DIR6—P0.6 P0DIR7—P0.7 P0.7 仅在 32 引脚 5 mm×5 mm 和 48 引脚 7 mm×7 mm 封装中有效

表 3.1.2 P1DIR 寄存器(地址—0x94；复位值—0xFF)

位	名称	R/W	功能
[7:0]	dir	R/W	确定引脚端 P1.0～P1.7 方向。输出 dir=0，输入 dir=1 P1DIR0—P1.0 P1DIR1—P1.1 P1DIR2—P1.2 P1DIR3—P1.3 P1DIR4—P1.4 P1DIR5—P1.5 P1DIR6—P1.6 P1DIR7—P1.7 I/O 端口 1 仅在 32 引脚 5 mm×5 mm 和 48 引脚 7 mm×7 mm 封装中有效 P1.7 仅在 48 引脚 7 mm×7 mm 封装中有效

表 3.1.3 P2DIR 寄存器(地址—0x95；复位值—0xFF)

位	名称	R/W	功能
[7:0]	dir	R/W	确定引脚端 P2.0～P2.7 方向。输出 dir=0，输入 dir=1 P2DIR0—P2.0 P2DIR1—P2.1 P2DIR2—P2.2 P2DIR3—P2.3 P2DIR4—P2.4 P2DIR5—P2.5 P2DIR6—P2.6 P2DIR7—P2.7 I/O 端口 2 仅在 48 引脚 7 mm×7 mm 封装中有效

表 3.1.4 P3DIR 寄存器(地址—0x96；复位值—0xFF)

位	名称	R/W	功能
[7:0]	dir	R/W	确定引脚端 P3.0～P3.7 方向。输出 dir=0，输入 dir=1 P3DIR0—P3.0 P3DIR1—P3.1 P3DIR2—P3.2 P3DIR3—P3.3 P3DIR4—P3.4 P3DIR5—P3.5 P3DIR6—P3.6 P3DIR7—保留 I/O 端口 3 仅在 48 引脚 7 mm×7 mm 封装中有效

2. PxCON 寄存器

引脚的输入/输出选择在 PxCON 寄存器中配置。PxCON 寄存器必须每个引脚写一次（PxCON 寄存器的写操作配置所选择端口的每一个引脚）。为了读取一个引脚当前是选择了输出还是输入，首先需要执行一个写操作来获取目的位地址和选项类型（输入或输出）。例如，要读出 P0.5 的输出模式，写以下值到 P0CON：位地址（bitAddr）值 3'b101，读地址（readAddr）值 1，inOut 值 0（输出）。然后，读出 P0CON，在读出数据的位[7:5]后即可发现 P0.5 的输出模式。PxCON 寄存器功能如表 3.1.5～表 3.1.8 所列。

表 3.1.5 P0CON 寄存器（地址—0x9E；复位值—0x00）

位	名称	R/W	功能
[7:5]	pinMode	R/W	设置 P0.0～P0.7 的输入/输出功能模式。 写操作：写入所希望的引脚功能模式。inOut 字段确定写入的是输入或输出模式，bitAddr 字段确定哪一个引脚端受到影响。 输出模式使用位[7:5]： 　　3'b000　数字输出缓冲器具有标准驱动能力； 　　3'b011　数字输出缓冲器具有高的驱动能力 　　（所有其他值组合都是禁止的）。 输入模式使用位[6:5]： 　　2'b00　数字输入缓冲器导通，无上拉/下拉电阻； 　　2'b01　数字输入缓冲器导通，连接下拉电阻； 　　2'b10　数字输入缓冲器导通，连接上拉电阻； 　　2'b11　数字输入缓冲器关闭。 读操作：读出引脚端的当前功能模式。inOut 字段确定读出的是输入或输出模式，bitAddr 字段指示哪一个引脚端被选择
[4]	inOut	W	该位指示当前写入操作所涉及的引脚端的输入或输出配置。 inOut=0—工作在输出配置；inOut=1—工作在输入配置
[3]	readAddr	W	如果该位置位，则当前写入操作的目的是为后面的读操作提供位地址。因此，bitAddr 字段的值被保存。inOut 字段值的也得以保存，确定输入或输出模式的读取。当设置 readAddr 时，pinMode 字段被忽略。 如果此位未置位，则被寻址引脚端的引脚模式被 pinMode 字段的值更新。inOut 字段确定输入或输出模式的更新
[2:0]	bitAddr	W	如果 readAddr 位置位，则 bitAddr 字段的值被存储。对于后续的读 P0CON 的操作，pinMode 将按照以下参数进行操作。 　　　　　　　　7 mm×7 mm　　5 mm×5 mm　　4 mm×4 mm bitAddr=3'b000　　P0.0　　　　P0.0　　　　P0.0 bitAddr=3'b001　　P0.1　　　　P0.1　　　　P0.1 bitAddr=3'b010　　P0.2　　　　P0.2　　　　P0.2 bitAddr=3'b011　　P0.3　　　　P0.3　　　　P0.3 bitAddr=3'b100　　P0.4　　　　P0.4　　　　P0.4 bitAddr=3'b101　　P0.5　　　　P0.5　　　　P0.5 bitAddr=3'b110　　P0.6　　　　P0.6　　　　P0.6 bitAddr=3'b111　　P0.7　　　　P0.7　　　　保留

第 3 章　nRF24LE1 的接口与应用

表 3.1.6　P1CON 寄存器(地址—0x9F;复位值—0x00)

位	名称	R/W	功能
[7:5]	pinMode	R/W	设置 P1.0~P1.7 的输入/输出功能模式 写操作：写入所希望的引脚功能模式。inOut 字段确定写入的是输入或输出模式，bitAddr 字段确定哪一个引脚端受到影响。 输出模式使用位[7:5]： 　　3'b000　数字输出缓冲器具有标准驱动能力； 　　3'b011　数字输出缓冲器具有高的驱动能力 （所有其他值组合都是禁止的）。 输入模式使用位[6:5]： 　　2'b00　数字输入缓冲器导通,无上拉/下拉电阻； 　　2'b01　数字输入缓冲器导通,连接下拉电阻； 　　2'b10　数字输入缓冲器导通,连接上拉电阻； 　　2'b11　数字输入缓冲器关闭。 读操作：读出引脚端的当前功能模式。inOut 字段确定读出的是输入或输出模式，bitAddr 字段指示哪一个引脚端被选择
[4]	inOut	W	该位指示当前写入操作所涉及的引脚端的输入或输出配置。 inOut=0—工作在输出配置；inOut=1—工作在输入配置
[3]	readAddr	W	如果该位置位,则当前写入操作的目的是为后面的读操作提供位地址。因此,bitAddr 字段的值被保存。inOut 字段值的也得以保存,确定输入或输出模式的读取。当设置 readAddr 时,pinMode 字段被忽略。 如果此位未置位,则被寻址引脚端的引脚模式被 pinMode 字段的值更新。inOut 字段确定输入或输出模式的更新
[2:0]	bitAddr	W	如果 readAddr 位置位,则 bitAddr 字段的值被存储。对于后续的读 P1CON 的操作,pinMode 将按照以下参数进行操作。 　　　　　　　　　　7 mm×7 mm　　5 mm×5 mm　　4 mm×4 mm bitAddr=3'b000　　P0.0　　　　　P0.0　　　　　保留 bitAddr=3'b001　　P0.1　　　　　P0.1　　　　　保留 bitAddr=3'b010　　P0.2　　　　　P0.2　　　　　保留 bitAddr=3'b011　　P0.3　　　　　P0.3　　　　　保留 bitAddr=3'b100　　P0.4　　　　　P0.4　　　　　保留 bitAddr=3'b101　　P0.5　　　　　P0.5　　　　　保留 bitAddr=3'b110　　P0.6　　　　　P0.6　　　　　保留 bitAddr=3'b111　　P0.7　　　　　保留　　　　　保留

表 3.1.7 P2CON 寄存器(地址—0x97;复位值—0x00)

位	名称	R/W	功能
[7:5]	pinMode	R/W	设置 P2.0～P2.7 的输入/输出功能模式 写操作：写入所希望的引脚功能模式。inOut 字段确定写入的是输入或输出模式，bitAddr 字段确定哪一个引脚端受到影响。 输出模式使用位[7:5]： 　　3'b000　数字输出缓冲器具有标准驱动能力； 　　3'b011　数字输出缓冲器具有高的驱动能力 （所有其他值组合都是禁止的）。 输入模式使用位[6:5]： 　　2'b00　数字输入缓冲器导通，无上拉/下拉电阻； 　　2'b01　数字输入缓冲器导通，连接下拉电阻； 　　2'b10　数字输入缓冲器导通，连接上拉电阻； 　　2'b11　数字输入缓冲器关闭。 读操作：读出引脚端的当前功能模式。inOut 字段确定读出的是输入或输出模式，bitAddr 字段指示哪一个引脚端被选择
[4]	inOut	W	该位指示当前写入操作所涉及的引脚端的输入或输出配置。 inOut=0—工作在输出配置；inOut=1—工作在输入配置
[3]	readAddr	W	如果该位置位，则当前写入操作的目的是为后面的读操作提供位地址。因此，bitAddr 字段的值被保存。inOut 字段值的也得以保存，确定输入或输出模式的读取。当设置 readAddr 时，pinMode 字段被忽略。 如果此位未置位，则被寻址引脚端的引脚模式被 pinMode 字段的值更新。inOut 字段确定输入或输出模式的更新
[2:0]	bitAddr	W	如果 readAddr 位置位，则 bitAddr 字段的值被存储。对于后续的读 P2CON 的操作，pinMode 将按照以下参数进行操作。 　　　　　　　　　7 mm×7 mm　　5 mm×5 mm　　4 mm×4 mm bitAddr=3'b000　　P2.0　　　　　保留　　　　　保留 bitAddr=3'b001　　P2.1　　　　　保留　　　　　保留 bitAddr=3'b010　　P2.2　　　　　保留　　　　　保留 bitAddr=3'b011　　P2.3　　　　　保留　　　　　保留 bitAddr=3'b100　　P2.4　　　　　保留　　　　　保留 bitAddr=3'b101　　P2.5　　　　　保留　　　　　保留 bitAddr=3'b110　　P2.6　　　　　保留　　　　　保留 bitAddr=3'b111　　P2.7　　　　　保留　　　　　保留

表 3.1.8 P3CON 寄存器(地址—0x8F;复位值—0x00)

位	名称	R/W	功能
[7:5]	pinMode	R/W	设置 P3.0~P3.7 的输入/输出功能模式 写操作:写入所希望的引脚功能模式。inOut 字段确定写入的是输入或输出模式, 　　　　bitAddr 字段确定哪一个引脚端受到影响。 输出模式使用位[7:5]: 　　3'b000　数字输出缓冲器具有标准驱动能力; 　　3'b011　数字输出缓冲器具有高的驱动能力 　　(所有其他值组合都是禁止的)。 输入模式使用位[6:5]: 　　2'b00　数字输入缓冲器导通,无上拉/下拉电阻; 　　2'b01　数字输入缓冲器导通,连接下拉电阻; 　　2'b10　数字输入缓冲器导通,连接上拉电阻; 　　2'b11　数字输入缓冲器关闭。 读操作:读出引脚端的当前功能模式。inOut 字段确定读出的是输入或输出模式, 　　　　bitAddr 字段指示哪一个引脚端被选择
[4]	inOut	W	该位指示当前写入操作所涉及的引脚端的输入或输出配置。 inOut=0—工作在输出配置;inOut=1—工作在输入配置
[3]	readAddr	W	如果该位置位,则当前写入操作的目的是为后面的读操作提供位地址。因此,bitAddr 字段的值被保存。inOut 字段值的也得以保存,确定输入或输出模式的读取。当设置 readAddr 时,pinMode 字段被忽略。 如果此位未置位,则被寻址引脚端的引脚模式被 pinMode 字段的值更新。inOut 字段确定输入或输出模式的更新
[2:0]	bitAddr	W	如果 readAddr 位置位,则 bitAddr 字段的值被存储。对于后续的读 P3CON 的操作,pinMode 将按照以下参数进行操作。 　　　　　　　　　　　7 mm×7 mm　　5 mm×5 mm　　4 mm×4 mm bitAddr=3'b000　　P3.0　　　　　保留　　　　　保留 bitAddr=3'b001　　P3.1　　　　　保留　　　　　保留 bitAddr=3'b010　　P3.2　　　　　保留　　　　　保留 bitAddr=3'b011　　P3.3　　　　　保留　　　　　保留 bitAddr=3'b100　　P3.4　　　　　保留　　　　　保留 bitAddr=3'b101　　P3.5　　　　　保留　　　　　保留 bitAddr=3'b110　　P3.6　　　　　保留　　　　　保留 bitAddr=3'b111　　保留　　　　　保留　　　　　保留

当 I/O 端口作为 MCU GPIO 使用时,引脚值的读和控制由 MCU 端口寄存器 P3~P0 控制。P3~P0 寄存器如表 3.1.9 所列。有多少 I/O 端口可以选择使用取决于 nRF24LEl 的封装形式。

表 3.1.9 P3～P0 寄存器

地 址	名 称	位	复位值	类 型	说 明
0xB0	P3	[7:0]	0xFF	R/W	通道 3 数值
0xA0	P2	[7:0]	0xFF	R/W	通道 2 数值
0x90	P1	[7:0]	0xFF	R/W	通道 1 数值
0x80	P0	[7:0]	0xFF	R/W	通道 0 数值

3.1.3 GPIO 与按键和 LED 的连接电路

一个 nRF24LE1 的 GPIO 应用示例如图 3.1.1 所示,利用按键控制 LED 发光二极管发光。按键 KEY0、KEY1 和 KEY2 作为 nRF24LE1 的 GPIO 的输入设备,3 个 LED 发光二极管则作为 nRF24LE1 的 GPIO 的输出设备。nRF24LE1 的 I/O 引脚端与外设的连接如表 3.1.10 所列。

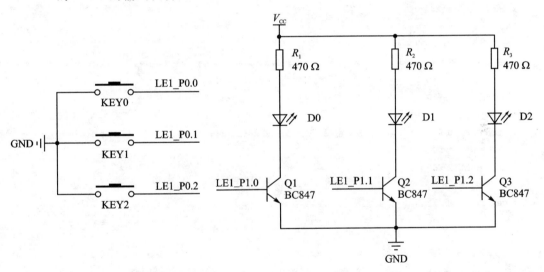

图 3.1.1 nRF24LE1 的 GPIO 与按键和 LED 的连接电路

表 3.1.10 nRF24LE1 的 I/O 引脚端与外设的连接

nRF24LE 的 I/O 引脚端	外设网络标号
P0.0	LE1_P0.0
P0.1	LE1_P0.1
P0.2	LE1_P0.2
P1.0	LE1_P1.0
P1.1	LE1_P1.1
P1.2	LE1_P1.2

第 3 章 nRF24LE1 的接口与应用

3.1.4 GPIO 示例程序流程图

nRF24LE1 的 GPIO 与按键和 LED 的连接电路示例程序的程序流程图如图 3.1.2 所示。

3.1.5 GPIO 示例程序

在本示例中，nRF24LE1 检测 I/O 引脚端的输入电平。如果检测到 P0.0 为低电平，那么就点亮 LED0；如果检测到 P0.1 为低电平，那么就点亮 LED1；如果检测到 P0.2 为低电平，那么就点亮 LED2。

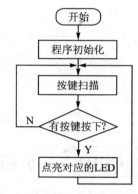

图 3.1.2 nRF24LE1 的 GPIO 应用示例程序流程图

示例程序的源代码如下：

```
#include "reg24le1.h"
/********************************************************
/宏定义 nRF24LE1 的一些 I/O 口作为按键和 LED
********************************************************/
#define led0 P10          /* 定义 LED0 为 nRF24LE1 的 P1.0 引脚 */
#define led1 P11          /* 这样做可以直观地操作对应引脚 */
#define led2 P12
#define KEY0 P00          /* 定义 KEY0 为 nRF24LE1 的 P0.0 引脚 */
#define KEY1 P01
#define KEY2 P02
#define false 0           /* 定义一些逻辑常量 */
#define true  1
/********************************************************
/函数声明
********************************************************/
void gpio_init(void);
void ledlight(unsigned char lnum);
unsigned char keyscan(void);
void shut(void);
/********************************************************
/函数名称：gpio_init()
/函数功能：初始化 nRF24LE1 的 I/O 口
/输入参数：无
/返回参数：无
********************************************************/
void gpio_init(void)
{
```

```
P0DIR = 0xFF;              /* P0 端口全部为输入 */
P1DIR = 0x00;              /* P1 端口全部为输出 */
P0CON = 0xD8;
P1CON = 0x00;
P1 = 0x00;
P0 = 0x00;
}
/******************************************************************
/函数名称: ledlight()
/函数功能: 点亮指定的 LED
/输入参数: lnum 为 LED 的编号
/返回参数: 无
******************************************************************/
void ledlight(unsigned char lnum)
{
if(lnum == false)          /* 如果没有按键按下,则返回 */
return ;
switch(lnum)               /* 如果有键按下,则根据键号点亮 LED */
    {
    case 1:shut();led0 = 1;break;
    case 2:shut();led1 = 1;break;
    case 3:shut();led2 = 1;break;
    default :break;
    }
}
/******************************************************************
/函数名称: keyscan()
/函数功能: 扫描按键按下情况
/输入参数: 无
/返回参数: 按下的按键的键值
******************************************************************/
unsigned char keyscan(void)
{
unsigned char flag = 0,num = 0;
P0CON = 0xD0;              /* 检测按键 0 */
if(!KEY0)
{
num = 1;                   /* 如果按下,按键号置为 1 */
flag = 1;
}
P0CON = 0xD1;              /* 检测按键 1 */
```

```c
    if(!KEY1)
    {
    flag = 1;
    num = 2;                    /*如果按下,按键号置为2*/
    }
    P0CON = 0xD2;               /*检测按键2*/
    if(!KEY2)
    {
        flag = 1;
        num = 3;                /*如果按下,按键号置为2*/
    }
    if(flag)                    /*如果有键按下,则这个标志为1,检测该标志判断按键状态*/
    return num;                 /*返回键号*/
    else
    return false;               /*返回false,表明无按键按下*/
}
/***************************************************************
/函数名称:shut()
/函数功能:关闭连接的所有LED灯
/输入参数:无
/返回参数:无
***************************************************************/
void shut(void)
{
led0 = 0;                       /*熄灭LED0*/
led1 = 0;
led2 = 0;
}


/***************************************************************
/主函数部分
***************************************************************/

void main(void)
{
gpio_init();                    /*I/O初始化*/
while(1)
  {
  ledlight(keyscan());          /*检测按键,并根据输入点亮对应的LED灯*/
  }
}
```

当上面的程序编译好并下载到 nRF24LE1 里后,复位 nRF24LE1。此时如果按下 KEY0,就会亮 LED0;按下 KEY1,就会点亮 LED1;按下 KEY2,就点会亮 LED2。这里的按键和 LED 都是使用了 nRF24LE1 的 GPIO 功能。

3.2 串行外设接口 SPI

3.2.1 SPI 结构与功能

nRF24LE1 具有一个双缓冲 FIFO 结构的 SPI(Serial Peripheral Interface,串行外设接口)接口,全双工工作,能够配置为工作在 4 种 SPI 模式(SPI 模式 0~3),默认工作模式为 0。SPI 接口连接到器件的以下引脚:MMISO、MMOSI、MSCK、SCSN、SMISO、SMOSI 和 SSCK。在 xMISO/xMOSI 上可配置数据顺序。SPI 主模式将不会产生任何片选信号(CSN)。编程者需要使用其他的可编程数字 I/O 产生片选信号,用来选择一个或多个外部 SPI 从设。具有 4 个 (主模式)和 3 个(从模式)中断源。

SPI 主模式示意图如图 3.2.1 所示,SPI 从模式示意图如图 3.2.2 所示。

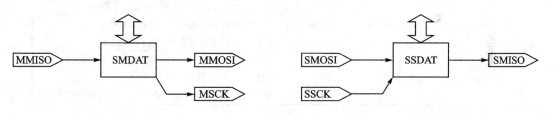

图 3.2.1　SPI 主模式示意图　　　　　　图 3.2.2　SPI 从模式示意图

3.2.2 SPI 主模式寄存器

SPI 主模式寄存器包含 SPIMCON0、SPIMCON1、SPIMSTAT 和 SPIMDAT,这些寄存器控制 SPI 的主模式。

1. SPIMCON0

SPI 主模式配置寄存器 0 SPIMCON0 功能如表 3.2.1 所列,SPIMCON0 的地址为 0xFC,位[6:0],复位值为 0x02。

SPI 主模式通过寄存器 SPIMCON0 和 SPIMCON1 来配置,通过设置 SPIMCON0 的位 [0]为 1 来使能。SPI 主模式支持全部 4 种 SPI 模式,由 SPIMCON0 的位[2]和位[1]来选择。MMISO/MMOSI 上的数据顺序由 SPIMCON0 的位[3]来设定。MSCK 可以设定为 6 种 MCU 时钟频率(1/2~1/64)之一,由 SPIMCON0 的位[6:4]来选择。

第3章 nRF24LE1 的接口与应用

表 3.2.1 SPIMCON0 功能

名称/助记符	位	复位值	R/W	功能
clockFrequency	[6:4]	010	R/W	MSCK 的频率。f_{ckCpu} 是 MCU 的时钟频率。 000：1/2 f_{ckCpu}。 001：1/4 f_{ckCpu}。 010：1/8 f_{ckCpu}。 011：1/16 f_{ckCpu}。 100：1/32 f_{ckCpu}。 101：1/64 f_{ckCpu}。 110：1/64 f_{ckCpu}。 111：1/64 f_{ckCpu}。
dataOrder	[3]	0	R/W	串行输出和输入的数据顺序（MMOSI 和 MMISO）。 1：LSB（低位）先，MSB（高位）后。 0：MSB（高位）先，LSB（低位）后
clockPolarity	[2]	0	R/W	与 SPIMCON0 的位[1]定义 SPI 主工作模式。 1：MSCK 是低电平有效。 0：MSCK 是高电平有效
clockPhase	[1]	0	R/W	与 SPIMCON0 的位[2]定义 SPI 主工作模式。 1：在 MSCK 的下降沿采样，在上升沿移位。 0：在 MSCK 的上升沿采样，在下降沿移位
spiMasterEnable	[0]	0	R/W	1：使能 SPI 主模式。时钟在 SPI 主模式内核运行。SPI 传输通过 8051 SFR 总线（TX）由 MCU 启动。 0：SPI 主模式不使能。时钟在 SPI 主模式内核不运行

2. SPIMCON1

SPI 主模式配置寄存器 1 SPIMCON1 功能如表 3.2.2 所列，SPIMCON1 的地址为 0xFD，位[3:0]，复位值为 0x0F。

表 3.2.2 SPIMCON1 功能

名称/助记符	位	复位值	R/W	功能
maskIrqRxFifoFull	[3]	1	R/W	1：当 RX FIFO 满时，不使能中断。 0：当 RX FIFO 满时，使能中断
maskIrqRxDataReady	[2]	1	R/W	1：当在 TX FIFO 有效时，不使能中断。 0：当在 TX FIFO 有效时，使能中断
maskIrqTxFifoEmpty	[1]	1	R/W	1：当在 TX FIFO 空时，不使能中断。 0：当在 TX FIFO 空时，使能中断
maskIrqTxFifoReady	[0]	1	R/W	1：当在 TX FIFO 位置可用时，不使能中断。 0：当在 TX FIFO 位置可用时，使能中断

除非 SPIMCON1 寄存器相应的控制位被屏蔽，SPI 有 4 个不同的中断源能够产生中断，SPIMSTAT 寄存器显示哪个中断源是有效的。

3. SPIMSTAT

SPI 主模式状态寄存器 SPIMSTAT 功能如表 3.2.3 所列，SPIMSTAT 的地址为 0xFE，位[3:0]，复位值为 0x03。

表 3.2.3　SPIMSTAT 功能

名称/助记符	位	复位值	R/W	功　　能
rxFifoFull	[3]	0	R	中断源。 1：RX FIFO 满。 0：RX FIFO 能够接收来自 SPI 更多的数据。 当产生原因消除时清除
rxDataReady	[2]	0	R	中断源。 1：在 RX FIFO 中的数据有效。 0：在 RX FIFO 中的数据无效。 当产生原因消除时清除
txFifoEmpty	[1]	1	R	中断源。 1：TX FIFO 空。 0：数据在 TX FIFO。 当产生原因消除时清除
txFifoEmpty	[0]	1	R	中断源。 1：Location available in 在 TX FIFO 中的位置有效。 0：TX FIFO 满。 当产生原因消除时清除

4. SPIMDAT

SPI 主模式数据寄存器 SPIMDAT 的地址为 0xFF，位[7:0]，复位值为 0x00，访问 TX（写）和 RX（读）FIFO 缓冲器，两字节深度。

通过 SPIMDAT 可以访问发射（写）和接收（读）FIFO 缓冲区。FIFO 是动态的，并可以依照状态位的状态进行填充，FIFO ready 标志意味着 FIFO 可以接收数据，Data ready 意味着可以提供数据。

3.2.3　SPI 从模式寄存器

SPI 从模式寄存器包含 SPISCON0、SPISSTAT 和 SPIMDAT，这些寄存器控制 SPI 的从模式。

1. SPISCON0

SPI 从模式配置寄存器 0 SPISCON0 功能如表 3.2.4 所列，SPISCON0 的地址为 0xBC，位[6:0]，复位值为 0x70。

SPI 从模式通过 SPISCON0 寄存器配置。通过设置 SPISCON0 的位[0]为 1 来使能。SPI 从模式支持 4 种 SPI 模式，由 SPISCON0 的位[2]和 SPISCON0 的位[1]来选择。SMISO/SMOSI 上的数据顺序由 SPISCON0 的位[3]来设定。

SPI 从模式有 3 个中断源，其中任何一个均可以被屏蔽。

表 3.2.4 SPISCON0 功能

名称/助记符	位	复位值	R/W	功能
maskIrqCsnHigh	[6]	1	R/W	1：当 SCSN 为高电平时，不使能中断。 0：当 SCSN 为高电平时，使能中断
maskIrqCsnLow	[5]	1	R/W	1：当 SCSN 为低电平时，不使能中断。 0：当 SCSN 为低电平时，使能中断
maskIrqSpiSlaveDone	[4]	1	R/W	1：当 SPI 从设完成 SPI 传输时，不使能中断。 0：当 SPI 从设完成 SPI 传输时，使能中断
dataOrder	[3]	0	R/W	在串行输入和输出上的数据顺序(SMOSI 和 SMISO)。 1：LSB(低位)先，MSB(高位)后。 0：MSB(高位)先，LSB(低位)后
clockPolarity	[2]	0	R/W	与 SPISCON0 的位[1]定义 SPI 从工作模式。 1：SSCK 是低电平有效。 0：SSCK 是高电平有效
clockPhase	[1]	0	R/W	与 SPIMCON0 的位[2]定义 SPI 从工作模式。 1：在 SSCK 的下降沿采样，在上升沿移位。 0：在 SSCK 的上升沿采样，在下降沿移位
spiSlaveEnable	[0]	0	R/W	1：使能 SPI 从模式。时钟在 SPI 从模式内核运行。SPI 传输被 SPI 主设(RX)启动。 0：SPI 从模式不使能。时钟在 SPI 从模式内核不运行

2. SPISSTAT

SPI 从模式状态寄存器 SPISSTAT 功能如表 3.2.5 所列，SPISSTAT 的地址为 0xBE，位[5:0]，复位值为 0x00。

当发生中断时，SPISSTAT 寄存器可以提供中断来源的信息。

3. SPISDAT

SPI 从模式数据寄存器 SPISDAT 的地址为 0xBF，位[7:0]，复位值为 0x00，访问 TX(写)/RX(读)FIFO 缓冲器。

表 3.2.5 SPISSTAT 功能

名称/助记符	位	复位值	R/W	功　能
csnHigh	[5]	0	R	中断源。 1：SCSN 的上升沿被检测。 0：SCSN 的上升沿未被检测。 读取后清除
csnLow	[4]	0	R	中断源。 1：SCSN 的下降沿被检测。 0：SCSN 的下降沿未被检测。 读取后清除
Reserved	[3:1]	0	R	保留
spiSlaveDone	[0]	0	R	中断源。 1：SPI 从设完成了 SPI 传输。 0：SPI 从设未完成 SPI 传输。 当产生原因消除时清除

通过 SPIMDAT 可以访问发射（写）/接收（读）FIFO 缓冲区。FIFO 是动态的，并可以依照状态位的状态进行填充，FIFO ready 标志意味着 FIFO 可以接收数据，Data ready 意味着可以提供数据。

SPISDAT 用来作为双向数据访问，在外部 SPI 主设的第一个时钟到来前，MCU 可以写入两字节到 SPISDAT，但在 SCSN 变为低电平前只能是一字节。SMOSI 上的数据从外部 SPI 主设传送到 SPI 从设的同时，从设的第一字节数据在 SMISO 上传送到外部主设。为了获得最大的数据吞吐率，当第一字节发送后，软件必须确保总有两字节在 TX 通道，即一个正在发送，而另外一个在等待传送。有两种方法可以达到此目的：

- 预装载两字节 TX 数据，然后每个 SPI 从模式中断发送一字节，直到传输完成。
- 预装载一字节发送数据，在第一个 SPI 从模式中断装载两字节，然后每个 SPI 从模式中断发送一字节，直到传输完成。对于一些采用最高 SSCK 频率的传输，采用这种方法，在第一字节和第二字节间需要暂停，给 MCU 时间来装载下一个发送数据。

3.2.4　SPI 时序

1. SPI 模式

4 种不同的 SPI 模式如表 3.2.6 所列。

2. SPI 时序

SPI 时序如图 3.2.3 所示。有关 SPI 时序参数的更多内容请登录 www.nordicsemi.com，查

第3章　nRF24LE1 的接口与应用

询 nRF24LE1 Ultra-low Power Wireless System On-Chip Solution Preliminary Product Specification v1.6。

表 3.2.6　4 种不同的 SPI 模式

SPI 模式	时钟极性	时钟相位	时钟移位沿		时钟采样沿	
0	0	0	尾随	下降沿	前端	上升沿
1	0	1	前端	上升沿	尾随	下降沿
2	1	0	尾随	上升沿	前端	下降沿
3	1	1	前端	下降沿	尾随	上升沿

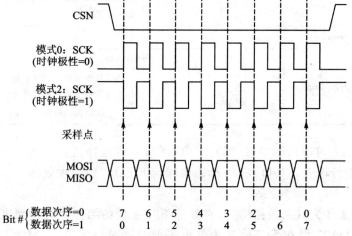

(a) SPI模式0和2：时钟相位=0，传送一字节

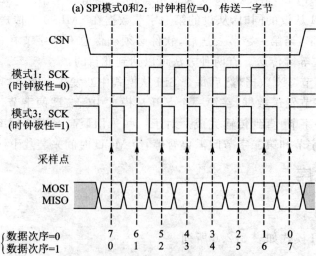

(b) SPI模式1和3：时钟相位=1，传送一字节

图 3.2.3　SPI 时序

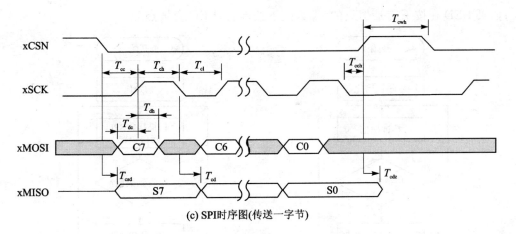

(c) SPI时序图(传送一字节)

图 3.2.3　SPI 时序(续)

3.2.5　SPI 主设与 SPI 从设之间的互联

SPI 总线包含有 xMISO、xMOSI、xSCK 和 xCSN 四条线,其中两条数据线,一条时钟线,一条片选线。nRF24LE1 的 SPI 总线接口,其中片选线必须采用一个 GPIO 来充当。在本示例程序中,将一个 nRF24LE1 模块设置为 SPI 主模式,另一个 nRF24LE1 模块设置为从模式,两个 nRF24LE1 模块的连接如表 3.2.7 所列,SPI 主设与 SPI 从设之间的互联示意图如图 3.2.4 所示。

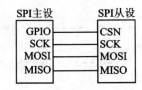

图 3.2.4　SPI 主设与 SPI 从设之间互联示意图

表 3.2.7　SPI 主设与 SPI 从设的 SPI 接口连接

主 nRF24LE1		从 nRF24LE1	
引脚	功能	引脚	功能
P1.6	MMISO	P1.0	SMISO
P1.5	MMOSI	P0.7	SMOSI
P1.4	MSCK	P0.5	SSCK
P1.3	MCSN	P1.1	SCSN

3.2.6　SPI 示例程序流程图

SPI 示例程序流程图如图 3.2.5 所示。在本示例中,使用了 SPI 总线接口的主模式和从模式。两个 nRF24LE1,其中一个 nRF24LE1 模块设置为 SPI 主模式,另一个 nRF24LE1 模块设置为从模式。主模式 nRF24LE1 通过 SPI 接口发出数据,从模式 nRF24LE1 通过 SPI 接口接收数据,并且将接收的数据通过一个 LED 的熄灭和点亮来指示传输的内容。主模式 nRF24LE1 发送的是一个按键状态,如果按键按下之前,LED 是亮的,那么按下的按键后 LED

熄灭；如果 LED 在按下前是熄灭的，那么按下按键后，LED 会被点亮。

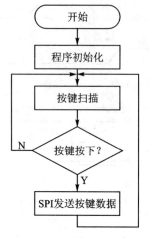
(a) nRF24LE1 主模式的程序流程图

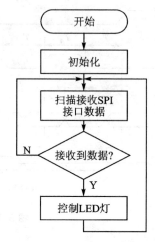
(b) nRF24LE1 从模式的程序流程图

图 3.2.5　SPI 示例程序流程图

3.2.7　SPI 示例程序

1. SPI 示例程序的宏定义

SPI 示例程序的宏定义如下：

```
#include "reg24le1.h"
/************************************************************
/主 SPI 接口状态的宏定义
*************************************************************/
#define txFifoReady    (SPIMSTAT&0x01)
#define rxDataReady    (SPIMSTAT&0x04)
#define txFifoEmpty    (SPIMSTAT&0x02)
#define rxFifoFull     (SPIMSTAT&0x08)
/************************************************************
/从 SPI 接口状态的宏定义
*************************************************************/
#define csnHigh        (SPISSTAT&0x20)
#define csnLow         (SPISSTAT&0x10)
#define spiSlaveDone   (SPISSTAT&0x01)
#define LED P12
#define CSN P13
/************************************************************
```

```
/函数声明
*****************************************************************/
void puts(char * str);
void putch(char ch);
void delay(unsigned int dx);
void uart();
void ioconfig();
unsigned char getdat();
unsigned char transmit(unsigned char dat);
void sspiconfig();
void mspiconfig();
unsigned char keyscan();
```

2. SPI 示例程序的接口相关函数

SPI 示例程序的接口相关函数如下：

```
/*****************************************************************
/函数名称：mspiconfig()
/函数功能：初始化主 SPI 接口
/输入参数：无
/返回参数：无
******************************************************************/
void mspiconfig()                 /* SPI 的配置函数 */
{
SPIMCON0 = (0x01 << 4)|0x01;     /*配置 SPI 时钟为 1/4·ckMCU,
                                    高位先,低位后,使能 SPI */
SPIMCON1 = 0x0F;                 /* 禁止所有 SPI 中断 */
}
/*****************************************************************
/函数名称：sspiconfig()
/函数功能：配置从 SPI 的寄存器
/输入参数：无
/返回参数：无
******************************************************************/
void sspiconfig()
{
SPISCON0 = 0x01;                 /* 使能从 SPI */
}
/*****************************************************************
/函数名称：transmit()
```

/函数功能：主机发送一个数据
/输入参数：dat 为要发送的一字节数据
/返回参数：SPI 接口收到的数据
***/

```c
unsigned char transmit(unsigned char dat)
{
  unsigned da;
  CSN = 0;
  if(txFifoReady)                    /* 发送缓冲准备好就发送 */
  SPIMDAT = dat;
/*  else
  {
   while(! txFifoReady);
    SPIMDAT = dat;
  }
   if(rxDataReady)
   da = SPIMDAT      ;
   CSN = 1;        */
   delay(5);
   return da;
}
```

/**
/函数名称：getdat()
/函数功能：从机接收一个数据
/输入参数：无
/返回参数：SPI 接口收到的数据
***/

```c
unsigned char getdat()
{
unsigned char dat;
if(spiSlaveDone)                   /* 如果接收完成 */
dat = SPISDAT;
return dat;
}
```

/**
/函数名称：ioconfig()
/函数功能：配置 nRF24LE1 的相关 I/O 口的输入/输出
/输入参数：无
/返回参数：无

```
/****************************************************************/
void ioconfig()
{
    P1DIR &= 0xC3;              /* 配置 P1.2、P1.3、P1.4 和 P1.5 为输出 */
    P1DIR |= 0x40;              /* 配置 P1.6 为输入 */
    CSN = 1;
    P1DIR &= 0xFE;              /* 配置从 SPI 的 MISO 为输出 */
    P1DIR |= 0x02;              /* 配置 SCSN 为输入 */
    P0DIR |= 0x80;              /* 配置 MOSI 为输入 */
    P0DIR |= 0x20;              /* 配置 SCLK 为输入 */
    P0DIR |= 0x01;              /* 配置 P0.0 为按键输入 */
    P00 = 1;
    LED = 0;                    /* 初始化为低电平 */
}
```

3. SPI 示例程序的串口功能函数

SPI 示例程序的串口功能函数如下：

```
/****************************************************************
/函数名称：uart()
/函数功能：配置串口波特率为 38 400,8-n-1 传输格式
/输入参数：无
/返回参数：无
*****************************************************************/
void uart()
{
    CLKCTRL = 0x28;             /* 设置 16 MHz 时钟 */
    CLKLFCTRL = 0x01;           /* 设置 32.768 kHz 实时时钟 */
    P0DIR &= 0xF7;              /* 配置 P0.3(TXD)为输出 */
    P0DIR |= 0x10;              /* 配置 P0.4(RXD)为输入 */
    P0 |= 0x18;
    S0CON = 0x50;
    PCON |= 0x80;               /* 波特率倍增 */
    WDCON |= 0x80;              /* 选择内部波特率发生器 */
    S0RELL = 0xFB;
    S0RELH = 0x03;              /* 波特率设置成 38 400 */
}
/****************************************************************
/函数名称：delay()
```

```
/*函数功能：软件延时函数
/*输入参数：dx为软件延时的时间数
/*返回参数：无
******************************************************************/
void delay(unsigned int dx)
{
    int dj;
    for(;dx>0;dx--)
        for(dj=120;dj>0;dj--)
        {
            ;
        }
}

/******************************************************************
/*函数名称：putch()
/*函数功能：串口打印一个字符
/*输入参数：ch为串口发送的一个字符
/*返回参数：无
******************************************************************/
void putch(char ch)
{
    S0BUF = ch;
    while(!TI0);                    /*等待串口发送完成*/
    TI0 = 0;
}

/******************************************************************
/*函数名称：puts()
/*函数功能：串口发送一个字符串
/*输入参数：str为字符串指针
/*返回参数：无
******************************************************************/
void puts(char * str)
{
    while( * str != '\0')           /*如果没有到结束符,就连续发送*/
    {
        putch( * str++);            /*发送一个字符*/
    }
}
```

4. SPI 示例程序的按键扫描函数

SPI 示例程序的按键扫描函数如下：

```
/*****************************************************************
/函数名称: keyscan()
/函数功能: 判断扫描按键按下情况,并返回当前的按键状态值
/输入参数: 无
/返回参数: 按键的按下情况
*****************************************************************/
unsigned char keyscan()
{
static char f = 0;
P0CON = 0xD0;                  /* 设置 P0.0 为输入 */
if(!P00)                       /* 如果按键按下 */
{
delay(5);
if(!P00)
  {
  while(!P00);
  if(f)                        /* 按键状态值设置,如果是 1,则设置为 0,反之设置为 1 */
  f = 0;
  else
  f = 1;
  return (f);                  /* 返回按键状态 */
  },
}
 return f;
}
```

5. SPI 示例程序的主函数

SPI 示例程序的主函数如下：

```
/*****************************************************************
/主函数
*****************************************************************/

#define MASTER                 /* 主 SPI 和从 SPI 编译定义,程序采用条件编译 */
```

```c
#ifdef  MASTER              /*如果定义了主设*/
char flag = 0;              /*定义标志位*/
#endif                      /*结束本段条件编译*/
void main()
{
char sf = 0;
ioconfig();
uart();
#ifdef MASTER               /*如果定义了主设*/
mspiconfig();               /*配置主模式的SPI*/
#else
sspiconfig();               /*配置从模式SPI*/
#endif
delay(10);                  /*延时*/
puts("this is a spi test demo program...");/*打印提示信息*/
putch('\n');                /*换行*/
while(1)
{
#ifdef   MASTER             /*条件编译*/
flag = keyscan();           /*键盘扫描返回当前按键的状态*/
transmit(flag);             /*通过SPI传送数据*/
LED = flag;
#else
sf = getdat();              /*通过从SPI接收数据*/
LED = sf;                   /*控制LED灯显示*/
putch('0' + sf);            /*串口显示接收的按键状态*/
putch('\n');
#endif
}
}
```

程序在编译环境工具中进行编译时，应注意一定要编译两次，并且这两次的编译是不一样的。为了程序的整齐和简洁，采用了条件编译的方式来编译整个程序。当MASTER被定义时，编译出来的HEX是nRF24LE1主模式SPI的程序；当MASTER被屏蔽时，编译出来的HEX是nRF24LE1从模式SPI的程序。将两个HEX分别下载到主SPI模式的nRF24LE1和从SPI模式的nRF24LE1中。按照图3.2.4连接主SPI模式nRF24LE1和从SPI模式nRF24LE1，上电后就可以实现前面描述的功能。

3.3 UART

3.3.1 UART 结构与功能

nRF24LE1 的 MCU 系统可以配置一个标准的 8051 串口,在器件的两个引脚 UART/RXD 和 UART/TXD 上可以分别提供 RXD 和 TXD 两个串口信号。串口的时钟来自 MCU 的时钟 ckCpu。UART 引脚的方向必须在 PxDIR 寄存器中分别设置,RXD 设置为输入,TXD 设置为输出。

UART(异步串行通信接口)接口具有如下特性:
- 同步模式,固定速率;
- 8 位 UART 模式,可变速率;
- 9 位 UART 模式,可变速率;
- 9 位 UART 模式,固定速率;
- 附加波特率发生器。

注意:不推荐使用定时器 1 溢出作为波特率发生器。

UART 接口示意图如图 3.3.1 所示。

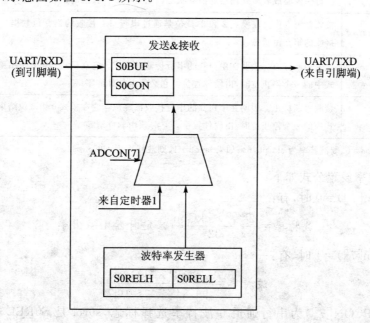

图 3.3.1 UART 接口示意图

3.3.2 UART 可编程寄存器

串口由 S0CON 寄存器控制,数据的传送通过读/写 S0BUF 寄存器完成。数据传送速率(波特率)利用 S0RELL、S0RELH 和 ADCON 寄存器来选择。

1. S0CON

S0CON(串口 0 控制寄存器)的控制功能如表 3.3.1 所列。

表 3.3.1 S0CON 寄存器功能(地址—0x98;复位值—0x00)

位	名 称	功 能
[7:6]	sm0:sm1	串行通道 0 模式选择。 0 0:Mode 0—移位寄存器,波特率为 ckCpu/12。 0 1:Mode 1—8 位 UART。 1 0:Mode 2—9 位 UART,波特率为 ckCpu/32 或 ckCpu/64①。 1 1:Mode 3—9 位 UART
[5]	sm20	多处理器通信使能
[4]	ren0	串行接收使能。1:使能串行通道 0
[3]	tb80	发送器第 8 位。在模式 2 和 3,此位是通过串行口 0 传输的第 9 位数据。该位的状态对应奇偶校验或者多处理器通信的状态。它是由软件控制的
[2]	rb80	接收器第 8 位。在模式 2 和 3,此位是通过串行口 0 接收的第 9 位数据。该位状态对应接收的第 9 位状态
[1]	ti0	发送中断标志。表明在串口 0 的串行传输已完成。它在模式 0 第 8 位结束时或在其他模式中的一个停止位时,由硬件置位。它必须由软件清零
[0]	ri0	接收中断标志。表明在串口 0 的串行接收已完成。它在模式 0 第 8 位结束时或在其他模式中的一个停止位时,由硬件置位。它必须由软件清零

① 如果 smod=0,则波特率为 ckCpu/64;如果 smod=1,则波特率为 ckCpu/32。

串口 0 波特率设置公式如下:

当 bd(adcon[7])=0 时,有:

$$波特率 = \frac{2^{SMOD} \times ckCpu}{32} \times 定时器 1 溢出率$$

当 bd(adcon[7])=1 时,有:

$$波特率 = \frac{2^{SMOD} \times ckCpu}{64 \times (2^{10} - s0rel)}$$

式中:SMOD(PCON[7])为串行通道 0 波特率选择标志,s0rel 是 S0REL 寄存器(s0relh,s0rell)的内容,bd(adcon[7])是 ADCON 寄存器的 MSB。波特率与寄存器的设置如表 3.3.2 所列。

表 3.3.2 波特率与寄存器的设置

波特率	Cclk	SMOD	s0rel	波特率	Cclk	SMOD	s0rel
600	16 MHz	1	0x00BF	9 600	16 MHz	1	0x03CC
1 200	16 MHz	1	0x025F	19 200	16 MHz	1	0x03E6
2 400	16 MHz	1	0x0330	38 400	16 MHz	1	0x03F3
4 800	16 MHz	1	0x0398				

2. S0BUF

S0BUF 的地址为 0x99，复位值为 0x00。写数据到寄存器 S0BUF（串口 0 数据缓冲器），将使数据移入串口输出缓冲器，并启动通过串口 0 发送；读寄存器 S0BUF，可以读出串行接收缓冲器所接收的数据。

3. S0RELH 和 S0RELL

S0RELH 和 S0RELL（串口 0 重载寄存器，地址为 0xBA 和 0xAA）用来作为串口 0 的波特率发生器。只使用了 10 位，其中 8 位在 S0RELL 寄存器，2 位在 S0RELH 寄存器。

4. ADCON

ADCON（串口 0 波特率选择寄存器）的最高位（MSB）用来设置串口 0 的波特率发生器，其功能如表 3.3.3 所列。

表 3.3.3 ADCON 寄存器功能

地 址	复位值	位	名 称	功 能
0xD8	0x00	[7]	bd	串口 0 波特率选择（模式 1 和 3）。当设置为 1 时，使用附加在内部的波特率发生器，否则使用 Timer1（定时器 1）的溢出
—	—	[6:0]		未使用

3.3.3 UART 示例程序流程图

nRF24LE1 UART 与计算机串口的连接电路请参考 1.2.4 小节。UART 示例程序流程图如图 3.3.2 所示。

本示例程序能够实现 nRF24LE1 的 UART 的接收和发送。利用 1.2.4 小节中图 1.2.10，可以将 nRF24LE1 的 UART 接口和计算机的串口连接起来。计算机通过串口发送给 nRF24LE1 数据，发送到 nRF24LE1 数据，通过 nRF24LE1 的串口回送给计算机。在计算机显示器上的接收窗口中，可以看见发送的数据显示。

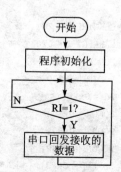

图 3.3.2 nRF24LE1 UART 示例程序流程图

3.3.4 UART 示例程序

UART 示例程序的源代码如下：

```c
#include "reg24le1.h"
/*******************************************************************
/宏定义常量
*******************************************************************/
#define true   1
#define false  0
#define LED0   P00
/*******************************************************************
/函数名称：io_config()
/函数功能：初始化本程序中使用到的 nRF24LE1 的 I/O 口
/输入参数：无
/返回参数：无
*******************************************************************/
void io_config(void)
{
    P0DIR &= 0xFE;
    LED0 = 0;
}
/*******************************************************************
/函数名称：uart_init()
/函数功能：初始化 nRF24LE1 的异步串口
/输入参数：无
/返回参数：无
*******************************************************************/
void uart_init(void)
{
    CLKCTRL = 0x28;          /* 设置 nRF24LE1 主时钟源 */
    CLKLFCTRL = 0x01;        /* 设置 32.768 kHz 实时时钟源 */
    P0DIR &= 0xF7;           /* 配置 P0.3(TXD)为输出 */
    P0DIR |= 0x10;           /* 配置 P0.4(RXD)为输入 */
    P0 |= 0x18;
    S0CON = 0x50;            /* 设置串口模式和接收使能 */
    PCON |= 0x80;            /* 波特率倍增 */
    WDCON |= 0x80;           /* 选择内部波特率发生器 */
    S0RELL = 0xF3;           /* 设置串口波特率为 38 400 */
    S0RELH = 0x03;
}
```

```
/*****************************************************************
/函数名称: send()
/函数功能: 利用串口发送一个字符
/输入参数: tmp 为一个待发送的字符
/返回参数: 无
*****************************************************************/
void send(unsigned char tmp)
{
S0BUF = tmp;
while(!TI0);
TI0 = 0;
}
/*****************************************************************
/函数名称: puts()
/函数功能: 通过串口发送一个字符串
/输入参数: s 为指向一个字符串的指针
/返回参数: 无
*****************************************************************/
void puts(unsigned char * s)
{
unsigned char * cs = s;
 while( * cs != '\0')
   {
    send( * cs ++);
   }
}
/*****************************************************************
/函数名称: getch()
/函数功能: 如果串口有接收到数据,就返回接收的数据,否则就返回空
/输入参数: 无
/返回参数: 串口接收到的数据,否则返回 false
*****************************************************************/
unsigned char getch(void)
{
unsigned char tmp;

if(RI0)                       /* 如果收到数据 */
{
    RI0 = 0;
    tmp = S0BUF;
    return tmp;               /* 返回收到的数据 */
```

```
    }
    else
    return false;
}
/*************************************************************
/函数名称：delayx()
/函数功能：实现软件延时
/输入参数：x为软件延时的时间数
/返回参数：无
**************************************************************/
void delayx(int x)
{
int da,db;
 for(da = 100;da>0;da -- )
    for(db = 0;db<x;db ++ )
      {
        ;
      }
}
code const unsigned char test_string[] = "this is a test string from le1 uart! ";
/*************************************************************
/主函数部分
**************************************************************/
void main(void)
{
unsigned char tmp = 0,flag = 0;
unsigned int num = 0;
io_config();                    /*I/O口初始化*/
uart_init();                    /*串口初始化函数*/
delayx(100);
puts(test_string);              /*显示程序信息*/
while(1)
  {
  if(RI0 == 1)                  /*如果收到串口数据*/
    {
      tmp = S0BUF;
      RI0 = 0;
      S0BUF = tmp;              /*返回接收的数据*/
      while(!TI0);
      TI0 = 0;
    }
```

第3章　nRF24LE1 的接口与应用

```
    num ++ ;
    if(num＜2000)
    flag = 1;
    else
    flag = 0;
    num % = 10000;
    LED0 = flag;
        }
}
/***************************************************************
/                           程序结束                              /
***************************************************************/
```

以上程序编译完成下载到 nRF24LE1 上运行之后，用串口线连接计算机串口和 nRF24LE1 的串口，通过计算机上的串口调试助手与 nRF24LE1 进行通信。通过串口调试助手发送出去的数据会被 nRF24LE1 接收后发送回来，显示在串口调试助手的接收窗口中，如图 3.3.3 所示。

图 3.3.3　nRF24LE1 与计算机的串口通信

3.4 2线接口

3.4.1 2线接口结构与功能

1. 2线接口特性

nRF24LE1有一个单缓冲的2线接口,可配置为主设或从设方式,支持主发送、主接收、从发送和从接收4种工作模式;用两种不同的速率来发送和接收数据,支持标准模式(100 kbps)和高速模式(400 kbps)两种波特率;支持广播;支持7位地址;支持从机停止串行时钟(SCL)。2线接口与I^2C兼容,不兼容CBUS总线。2线接口连接到器件的引脚端W2SDA和W2SCL。

2. 2线接口推荐使用条件

2线接口推荐使用条件如下:
- 2线接口使能位(W2CON0[0])必须在其他2线操作前的一个单独写周期内置1。
- 如果使用了时钟停止特性,则时钟停止位(W2CON0[6])必须在传送开始前置1。在时钟停止模式(clockStop模式),所有接收数据必须从W2DAT寄存器读出。这是为了避免阻塞2线总线所必需的。
- 更新maskIrq配置位(W2CON1[5])必须在传输开始前完成。
- 一旦1已经写入xStart位(W2CON0[4])或xStop位(W2CON0[5])后,使用者不应试图清除该位。

3.4.2 2线接口主设发送/接收

进入启动条件后将开始新的传送。这可以通过设置W2CON0[4]为1,或简单地写第一字节到W2DAT来实现。第一字节总是由主设发出。

1. 发送模式

进入发送模式,MCU必须写地址到从设,写欲与其通信的从机地址或通用呼叫地址(0x00)到W2DAT[7:1]中。方向位(direction bit)W2DAT[0]写0。这个字节传送到从设。如果没有屏蔽,将会在该字节最后一位的SCL上升沿产生一个中断。同时,来自从设的应答信号存储在W2CON1[1]中。2线接口可以读来自通过MCU发送的数据,字节传输可以采用与传输第一字节相同的程序继续进行。

再次启动时,在写一个新的从机地址和方向位到W2DAT前,MCU必须置位W2CON0[4]。为了停止传输,必须在写最后一个发送字节到W2DAT后,写一个1到W2CON0[5]。

启动和停止的优先级低于待发送数据,W2CON0[4]和W2CON0[5]位可在写入最后一个发送字节后立即置位。如果两个位同时置位,则首先发送停止条件。

2. 接收模式

进入接收模式,MCU 必须写地址到从设,写欲与其通信的从设地址写入到 W2DAT[7:1],方向位 W2DAT[0]写 1。这个字节发送到从设。如果没有屏蔽,将会在该字节最后一位的 SCL 上升沿产生一个中断。同时,来自从设的应答信号存储在 W2CON1[1]中。然后,2 线接口释放总线,准备接收来自从设的数据。每收到一字节,如果没有屏蔽,每当收到最后一位时,产生中断,优先发送应答信号给从设。应答信号也存储在 W2CON1[1]中。

为重复传输或停止传输,在接收到第二个到最后一个来自从设的字节后,MCU 必须置位 W2CON0[5]。这使得 2 线主设在最后一字节后发送一个非应答信号,用来强迫从设释放总线控制。当接收完最后一字节后,主机可通过向 W2DAT 写入新的从设地址和方向位来启动一次新的传输。

3.4.3　2 线接口从设发送/接收

当从设检测到一个启动条件后,将进入接收模式,等待来自主设的第一字节。当第一字节完成时,从设将 W2DAT[7]到 W2DAT[1]位与 W2SADR 地址(或通用广播地址 0x00)进行比较,以确定是否应该回应。如果地址相同,则 W2DAT[0]将决定从设是保持在接收模式(0)还是进入发送模式(1)。当以下条件发生时,从设将向 MCU 产生中断请求:

① 在启动后,地址相匹配;
② 每收到该字节(接收模式)或发送一字节(发送模式)后;
③ 检测到一个停止条件。

所有中断均可通过配置来屏蔽。如果从设的 MCU 处理数据的速度不够快,则可通过在字节之间将 W2CON0[6]置 1 来延迟传输。在发送模式,这将强制 SCL 在传送后为低电平,直到 MCU 写一个新的数据到 W2DAT。在接收模式,SCL 在接收后保持为低电平,直到 MCU 读取一个新的数据。

新的发送数据必须由 MCU 在 SCL 的下一个下降沿前写入 W2DAT。新的接收数据必须在 SCL 的下一个下降沿前由 MCU 从 W2DAT 中读出。

3.4.4　2 线接口时序

2 线接口时序图如图 3.4.1～图 3.4.3 所示。有关时序参数请登录 www.nordicsemi.com,查询 nRF24LE1 Ultra-low Power Wireless System On-Chip Solution Preliminary Product Specification v1.6。

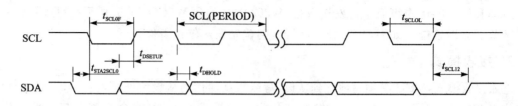

图 3.4.1　SCL/SDA 时序

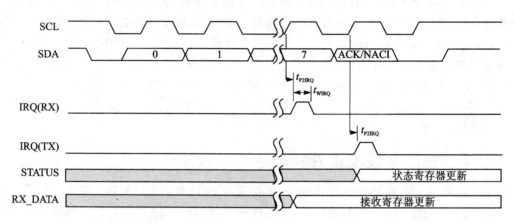

图 3.4.2　对 MCU 的中断请求时序

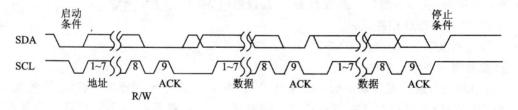

图 3.4.3　完整的数据传输

3.4.5　2 线接口特殊功能寄存器

MCU 通过特殊功能寄存器控制 2 线接口。

1. W2CON0

2 线接口配置寄存器 W2CON0 功能如表 3.4.1 所列,地址为 0xE2,位[7:0];复位值为 0x80。2 线接口通过 W2CON0[0]置位使能,W2CON0[1]决定是作为主设还是从设,波特率在 W2CON0[3:2]中设置。

注意:在标准模式,2 线接口通信需要至少 4 MHz 的系统时钟,在高速模式则需要 8 MHz 以上的系统时钟。

第 3 章　nRF24LE1 的接口与应用

表 3.4.1　2 线接口配置寄存器 W2CON0 功能

名称/助记符	位	复位值	R/W	功能
broadcastEnable	[7]	1	R/W	只有从设。 1：响应通用调用地址(0x00)，以及在 WIRE2ADR 定义的地址。 0：只响应在 WIRE2ADR 定义的地址
clo ckStop	[6]	0	R/W	只有从设。 1：在字节传送之间由从设保持 SCL 为低电平。MCU 接收数据或写发送数据。在发送模式，发送数据已写入 W2DAT 后的 t_{REL} 释放 SCL。在标准模式和快速模式，$t_{REL}=1\,400$ ns。在高速模式，$t_{REL}=5\times T_{ckCPU}$。在接收模式，从 W2DAT 读取数据后，立即释放 SCL。 注意：在传送开始之前更新此位。 0：2 线从设不改变时钟
xStop	[5]	0	R/W	只有主设。 1：发送停止条件。 　　RX 模式：在正在进行的字节接收完成后。 　　TX 模式：在任何等待的数据被发送。 注意：不要试图写 0 给它清除停止位。 0：没有停止条件被发送。 当清除时，发送停止条件。 只有从设。 1：当检测到停止条件时禁止中断。 0：当检测到停止条件时使能中断
xStart	[4]	0	R/W	只有主设。 1：在任一个代发送数据或者停止条件后，发送启动条件(重新启动)。 注意：不要试图写 0 给它清除启动位。 0：没有启动条件被发送。 当清除时，发送启动条件(重新启动) 只有从设。 1：在地址匹配时禁止中断。 0：在地址匹配时使能中断
clockFrequency	[3:2]	00	R/W	SCL 的频率。 00：空闲。 01：100 kHz (标准模式)，要求系统时钟频率至少为 4 MHz。 10：400 kHz (快速模式)，要求系统时钟频率至少为 8 MHz。 11：保留

第3章 nRF24LE1 的接口与应用

续表 3.4.1

名称/助记符	位	复位值	R/W	功能
masterSelect	[1]	0	R/W	1：选择主设模式。 0：选择从设模式
wire2Enable	[0]	0	R/W	1：使能2线接口。在2线内核的时钟运行。一个2线传输可以通过 MCU 的 8051 SFR 总线（TX）启动。 注意：在任何其他2线配置写操作之前，该位必须置位。 0：2线接口被禁用。在2线内核的时钟停止运行

2. W2CON1

2线接口配置寄存器/状态寄存器 W2CON1 功能如表 3.4.2 所列，地址为 0xE1，位 [5:0]，复位值为 0x00。

表 3.4.2 2线接口配置寄存器 W2CON1 功能

名称/助记符	位	复位值	R/W	功能
maskIrq	[5]	0	R/W	1：禁止所有中断。 0：使能所有中断。 注意：必须在任一个传输开始之前更新此位
broadcast	[4]	—	R	只有从设。 1：最后接收的地址是广播地址(0x00)。 0：最后接收的地址不是广播地址(0x00)。 读 W2CON1 时被清除
stop	[3]	—	R	只有从设。 1：中断由停止条件引起。 0：没有中断由停止条件引起。 读 W2CON1 时被清除
addressMatch	[2]	—	R	只有从设。 1：中断由地址匹配引起。 0：没有中断被地址匹配引起。 读 W2CON1 时被清除
ack_n	[1]	—	R	1：没有应答信号(NACK)。 0：有应答信号(ACK)。 该位确认2线接口已在最后发送后收到。 读 W2CON1 时被清除
dataReady	[0]	—	R	1：中断由发送/接收字节引起。 0：没有中断由发送/接收字节引起。 读 W2CON1 时被清除

3. W2SADR

2线接口从设地址寄存器 W2SADR,地址为 0xD9,位[6:0],复位值为 0x00。在 2 线接口从模式的地址。

4. W2DAT

2线接口数据寄存器 W2DAT,地址为 0xDA,位[7:0],复位值为 0x00。访问 TX(写)和 RX(读)缓冲器,一字节深度。

3.4.6　2 线接口应用示例电路

一个采用 nRF24LE1 2 线总线接口与使用 2 线总线的 EEPROM 芯片 CAT24WC02 连接电路如图 3.4.4 所示。SCL 连接到 nRF24LE1 引脚端 P0.4,SDA 连接到 nRF24LE1 引脚端 P0.5。

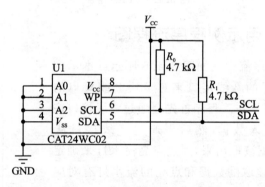

图 3.4.4　nRF24LE1 2 线总线与 CAT24WC02 的连接电路

电路中,EEPROM 芯片 CAT24WC02 作为 2 线从设,其地址是 0xA0。nRF24LE1 通过 2 线总线对该芯片进行读/写操作。

在程序中,首先将会在 EEPROM 芯片的前 4 个地址写入数据,然后通过读取这 4 个地址的值来点亮流水灯。如果读出的数据是 1,那么就点亮第 1 个 LED;如果读出的数据是 2,那么程序就会点亮第 2 个 LED;同理,如果读出的值是 3,就点亮第 3 个;读出的值是 4,就点亮第 4 个 LED。程序运行的过程中,EEPROM 只会在程序初始化阶段被写入数据,写入时在地址 0~4 写入的数据分别是 1~4,之后程序会利用定时器定时,每隔 1 s 从 EEPROM 中读取一个数。读取数据时,从地址 0 开始,每隔 1 s 地址会增加 1,如果读的地址累计到了 4,那么就重新从地址 0 开始读取,这样往返地执行下去。LED 电路连接如图 3.4.5 所示,nRF24LE1 的 I/O 口 P1.0~P1.4 依次连接图中的网络标号 LE1_P1.0~LE1_P1.4。

第 3 章 nRF24LE1 的接口与应用

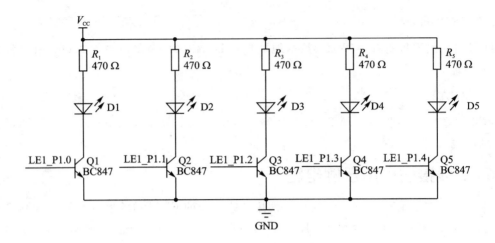

图 3.4.5 LED 电路

3.4.7 2 线接口应用示例程序流程图

2 线接口应用示例程序流程图如图 3.4.6 所示。本示例程序通过配置 nRF24LE1 的 2 线接口来对一块 2 线接口的 EEPROM 芯片进行读/写操作。程序首先会通过 2 线接口往 EEPROM 中指定地址写入数据,然后再通过 2 线接口从 EEPROM 中将之前的数据读取出来,通过点亮流水灯来确定数据的读/写正确与否。读取的数据和点亮的流水灯有对应关系,为了便于程序运行状态的监控,设置了一个状态 LED,用于程序的调试。

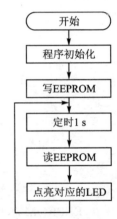

图 3.4.6 2 线接口应用示例程序流程图

3.4.8 2 线接口应用示例程序

1. 2 线接口应用示例程序的宏定义

2 线接口应用示例程序的宏定义如下:

```
/********************************************************
/说明部分:这是一个 nRF24LE1 的 2 线接口程序
/实现功能:读/写一块 2 线接口的 EEPROM,实现流水灯功能
/********************************************************/
#include "reg24le1.h"
/********************************************************
/LED 和 nRF24LE1 的 I/O 口关系的宏定义
*********************************************************/
```

```
#define LED0    P10
#define LED1    P11
#define LED2    P12
#define LED3    P13
#define LED4    P14
/*****************************************************************
/EEPROM 的地址宏定义
*****************************************************************/
#define SlaveAddr       0xA0
#define LimitAddr       0x04
#define StartAddr       0x00
#define EndAddr         0xFF
#define onetime         0x8235
/*****************************************************************
/逻辑宏定义
*****************************************************************/
#define true            1
#define false           0
/*****************************************************************
/宏定义 W2CON0 的各个位
*****************************************************************/
#define BROADCAST_ENABLE    7
#define CLOCK_STOP          6
#define X_STOP              5
#define X_START             4
#define CLOCK_FREQUENCY_1   3
#define CLOCK_FREQUENCY_0   2
#define MASTER_SELECT       1
#define WIRE_2_ENABLE       0
/*****************************************************************
/模式定义
*****************************************************************/
#define MasterMode      1
#define SlaveMode       0
#define UARTMOD         2
#define IICMOD          3
/*****************************************************************
/中断控制宏定义
*****************************************************************/
```

```c
#define Disableint()    do{EA = 0;}while(0)
#define Enableint()     do{EA = 1;}while(0)
/*****************************************************************
/类型重定义
*****************************************************************/
typedef unsigned int uint;
typedef unsigned char uchar;
/*****************************************************************/
enum LED {L0 = 1,L1,L2,L3 };
enum CLK {IDLE,CLK0,CLK1};
/*****************************************************************/
static uchar c_flag = 0;        /* 传输完成标志位 */
```

2. 2 线接口应用示例程序的常规功能函数

2 线接口应用示例程序的常规功能函数如下：

```c
/*****************************************************************
/函数名称：delay()
/函数功能：软件延时
/输入参数：x 为延时的时间数
/返回参数：无
*****************************************************************/
void delay(uint x)
{
uchar i;
    for(;x>0;x--)
      for(i = 120;i>0;i--)
        {
        }
}
/*****************************************************************
/函数名称：lightled()
/函数功能：根据指定的参数点亮对应的 LED
/输入参数：L 为 LED 编号
/返回参数：无
*****************************************************************/
void lightled(uchar L)
{
P1 &= 0xF0;                     /* 熄灭所有的 LED */
switch(L)
```

```
    {
        case L0 :LED0 = 1;break;
        case L1 :LED1 = 1;break;
        case L2 :LED2 = 1;break;
        case L3 :LED3 = 1;break;
        default :P1& = 0xF0;
    }
    return ;
}
/***************************************************************
/函数名称：Ldebug()
/函数功能：LED 闪烁功能
/输入参数：无
/返回参数：无
****************************************************************/
void Ldebug (void)
{
    LED4 = !LED4;
    delay(1000);
}
```

3. 2 线接口应用示例程序的操作基本函数

2 线接口应用示例程序的操作基本函数如下：

```
/***************************************************************
/函数名称：EnableWire()
/函数功能：2 线接口的使能或者关闭
/输入参数：flag 为 2 线接口使能或者关闭的标志量
/返回参数：无
****************************************************************/
void EnableWire(uchar flag)
{
    W2CON0 & = 0x7F;
if(flag == true)
    {
        W2CON0 |= BIT_0;            /*使能 2 线接口功能*/
    }
else
    {
```

```c
        W2CON0 &= 0xFE;              /*关闭2线接口功能*/
    }
}
/***************************************************************
/函数名称:WireClkSelect()
/函数功能:2线接口的工作频率设定
/输入参数:Clk为工作时钟选择参量,根据Clk的值设定2线接口时钟
/返回参数:无
***************************************************************/
void WireClkSelect(uchar Clk)
{
    switch(Clk)
    {
    case IDLE:W2CON0 &= 0xF3;break;
    case CLK0:W2CON0 = ((W2CON0&0xF3)|(CLK0 << CLOCK_FREQUENCY_0));break;
    case CLK1:W2CON0 = ((W2CON0&0xF3)|(CLK1 << CLOCK_FREQUENCY_0));break;
    default:break;
    }
    return ;
}
/***************************************************************
/函数名称:WireMode()
/函数功能:2线工作模式的设置
/输入参数:mode为2线接口工作模式量
/返回参数:无
***************************************************************/
void WireMode(uchar mode)
{
    if(mode == MasterMode)
    {
        W2CON0 |= BIT_1;
    }
    else
    {
        W2CON0 &= (~(BIT_1));
    }
}
/***************************************************************
/函数名称:xstart()
```

/*函数功能：2线重新发起新的一轮传输
/*输入参数：flag 为用于表示是否开始新的一轮传输
/*返回参数：无
**/
```c
void xstart(uchar flag)
{
    if(flag)
    {
    W2CON0 = W2CON0 | (1 << X_START);
    }
    else
    {
    W2CON0 = W2CON0 & ~(1 << X_START);
    }
}
```
/**
/*函数名称：xstop()
/*函数功能：用于暂停当前的2线传输
/*输入参数：flag 为决定是否要中断当前2线传输的一个标志
/*返回参数：无
**/
```c
void xstop(uchar flag)
{
    if(flag)
    {
      W2CON0 = W2CON0 | (1 << X_STOP);
    }
    else
    {
      W2CON0 = W2CON0 & ~(1 << X_STOP);
    }
}
```
/**
/*函数名称：IRQSet()
/*函数功能：2线传输中断设置
/*输入参数：flag 决定是否开启2线接口传输中断
/*返回参数：无
**/
```c
void IRQSet(uchar flag)
```

```c
{
    if(true == flag)
    {
        W2CON1 = ~(BIT_5);
        INTEXP = 0x04 ;              /* 开启 2 线完成中断 */
        T2CON = 0x40;                /* 选择上升沿触发 */
        IEN1 = 0x04;                 /* 使能 2 线完成中断 */
    }
    else
    {
        W2CON1 = (BIT_5);
    }
}
/****************************************************************
/函数名称：StatusGet()
/函数功能：读取 2 线状态寄存器的值
/输入参数：无
/返回参数：无
****************************************************************/
uchar StatusGet(void)
{
    return W2CON1;
}
/****************************************************************
/函数名称：Writedata()
/函数功能：往 2 线接口写入数据 dat
/输入参数：dat 为要通过 2 线接口发送的数据
/返回参数：无
****************************************************************/
void Writedata(uchar dat)
{
    W2DAT = dat;
}
/****************************************************************
/函数名称：Readdata()
/函数功能：读 2 线接口数据
/输入参数：无
/返回参数：从 2 线接口上接收到的数据
****************************************************************/
```

```
uchar Readdata(void)
{
return W2DAT;
}
/***************************************************************
/函数名称：WaitWireFree()
/函数功能：等待数据传输完成
/输入参数：无
/返回参数：无
***************************************************************/
void WaitWireFree(void)
{
while(!c_flag);              /*等待传输完成标志位置1*/
c_flag = 0;                  /*清除传输完成标志*/
}
```

4. 2 线接口应用示例程序的初始化函数

2 线接口应用示例程序的初始化函数如下：

```
/***************************************************************
/函数名称：Iocnfg()
/函数功能：初始化 nRF24LE1 的 I/O 口
/输入参数：无
/返回参数：无
***************************************************************/
void Iocnfg(void)
{
P1DIR &= 0xE0;
P0DIR &= 0xEF;
P04 = 1;
P1 &= 0xE0;
}
/***************************************************************
/函数名称：clkset()
/函数功能：nRF24LE1 的工作时钟设置
/输入参数：无
/返回参数：无
***************************************************************/
void clkset(void)
```

```
{
    CLKCTRL = 0x28;              /* XCOSC 16 MHz 主时钟源 */
    CLKLFCTRL = 0x01;            /* 32.768 kHz 实时时钟源 */
}
/***************************************************************
/函数名称：t0_init()
/函数功能：初始化定时器 0
/输入参数：无
/返回参数：无
****************************************************************/
void t0_init(void)
{
    TMOD = 0x01;
    TH0 = (65536 - onetime)/256;
    TL0 = (65536 - onetime) % 256;
    ET0 = 1;
    TR0 = 0;
}
/***************************************************************
/函数名称：IIC_init()
/函数功能：初始化 nRF24LE1 的 2 线接口
/输入参数：无
/返回参数：无
****************************************************************/
void IIC_init()
{
    EnableWire(true);            /* 使能 2 线接口 */
    WireClkSelect(CLK0);         /* 设置成 100 kHz */
    WireMode(MasterMode);        /* 设置成主模式 */
    IRQSet(true);                /* 使能中断 */
}
/***************************************************************
/函数名称：Uart_init()
/函数功能：初始化 nRF24LE1 异步串口功能
/输入参数：无
/返回参数：无
****************************************************************/
void Uart_init()
{
```

```
    PODIR &= 0xF7;          /*配置 P0.3(TXD)为输出*/
    PODIR |= 0x10;          /*配置 P0.4(RXD)为输入*/
    S0CON = 0x40;           /*设置串口的工作模式*/
    PCON |= 0x80;           /*波特率倍增*/
    WDCON |= 0x80;          /*选用内部波特率发生器*/
    S0RELH = 0x03;
    S0RELL = 0xF3;          /*设置波特率为 38 400*/
}
```

5.2 线接口应用示例程序的串口功能函数

2线接口应用示例程序的串口功能函数如下：

```
/***************************************************************
/函数名称：putch()
/函数功能：通过串口发送一个字符
/输入参数：Txdata 为串口待发送的字符数据
/返回参数：无
***************************************************************/
void putch(char Txdata)
{
S0BUF = Txdata;
while(!TI0);
TI0 = 0;
}
/***************************************************************
/函数名称：nextline()
/函数功能：利用串口发送一个换行命令
/输入参数：无
/返回参数：无
***************************************************************/
void nextline(void)
{
putch('\n');
}
/***************************************************************
/函数名称：puts()
/函数功能：串口发送一个字符串
/输入参数：str 为指向一个字符串的指针
/返回参数：无
```

```
***********************************************************/
void puts(char * str)
{
while( * str != '\0')
    {
    putch( * str ++ );
    }
}
/***********************************************************
/函数名称：RegUartDebug()
/函数功能：利用串口显示出指定寄存器的值
/输入参数：Reg 为 nRF24LE1 可以读取的寄存器
/返回参数：无
***********************************************************/
void RegUartDebug (uchar Reg)
{
static uchar num = 1;
nextline();
puts("REG");
putch('0' + num);
puts(" value: ");
putch((Reg/100) + '0');
putch((Reg % 100)/10 + '0');
putch((Reg % 10) + '0');
num ++ ;
if(4 == num)
        {
        num = 1;
        }
        delay(2);
nextline();
}
```

6. 2 线接口应用示例程序的读/写 EEPROM 函数

2 线接口应用示例程序的读/写 EEPROM 函数如下：

```
/***********************************************************
/函数名称：writebyte()
/函数功能：通过 nRF24LE1 的 2 线接口向 EEPROM 写一字节
```

/输入参数：Addr 为 EEPROM 中的地址，dat 为写入的数据
/返回参数：无
***/
```
void writebyte(uint Addr,uchar dat)
{
 xstart(true);
 Writedata(SlaveAddr);
 WaitWireFree();
 Writedata(Addr);
 WaitWireFree();
 Writedata(dat);
 WaitWireFree();
 xstop(true);
}
```
/***
/函数名称：WriteROM()
/函数功能：往 EEPROM 里连续写入一组数据
/输入参数：无
/返回参数：无
***/
```
void WriteROM()
{
 uchar wAddr = StartAddr;
 for(;wAddr<LimitAddr;wAddr++)
 {
  writebyte(wAddr,(wAddr+1));
 }
}
```
/***
/函数名称：readbyte()
/函数功能：通过 2 线接口从 EEPROM 任意有效地址读取一字节
/输入参数：Addr 为有效的 EEPROM 地址
/返回参数：从 Addr 地址读取的数据
***/
```
uchar readbyte(uint Addr)
{
 uchar Res = 0xFF;
 xstart(true);
 Writedata(SlaveAddr);
```

```
    WaitWireFree();
    Writedata(Addr);
    WaitWireFree();
    xstart(true);
    Writedata(SlaveAddr + 1);
    WaitWireFree();
    WaitWireFree();
    Res = Readdata();
    xstop(true);
    return Res;
}
/*******************************************************************
/函数名称：ReadROM()
/函数功能：从 EEPROM 里读取设置好的 EEPROM 地址的数据
/输入参数：无
/返回参数：EEPROM 中读取的结果
*******************************************************************/
uchar ReadROM(void)
{
static uint CurrentAddr = 0;
    uchar Res = 0xFF;
    Res = readbyte(CurrentAddr ++ );
    if(4 == CurrentAddr)CurrentAddr = 0;
return Res;
}
```

7. 2 线接口应用示例程序的中断服务函数

2 线接口应用示例程序的中断服务函数如下：

```
/*******************************************************************
/函数名称：T0ISR()
/函数功能：处理定时器 0 中断
/输入参数：无
/返回参数：无
*******************************************************************/
void T0ISR() interrupt INTERRUPT_TF0
{
static uchar Cnum = 0;
TR0 = 0;
```

```
    TH0 = (65536 - onetime)/256;
    TL0 = (65536 - onetime) % 256;
    Cnum ++ ;
    if(40 == Cnum)
     {
      Cnum = 0;
      lightled(ReadROM());
     }
    TR0 = 1;
}
/***************************************************************
/函数名称：WireISR()
/函数功能：处理2线传输完成中断
/输入参数：无
/返回参数：无
***************************************************************/
void WireISR() interrupt INTERRUPT_MSDONE
{
    c_flag = 1;
}
```

8. 2线接口应用示例程序的主函数

2线接口应用示例程序的主函数如下：

```
/***************************************************************
/主函数部分
***************************************************************/
void main()
{
    Disableint();              /*关闭所有的中断*/
    clkset();                  /*nRF24LE1 时钟设置*/
    Iocnfg();                  /*nRF24LE1 的 I/O 配置*/
    t0_init();                 /*初始化定时器 0*/
    IIC_init();                /*2线接口的初始化*/
    delay(100);
    Enableint();               /*中断使能*/
    WriteROM();                /*给 EEPROM 写入数据*/
    delay(100);
    TR0 = 1;                   /*启动定时器*/
    while(1)
```

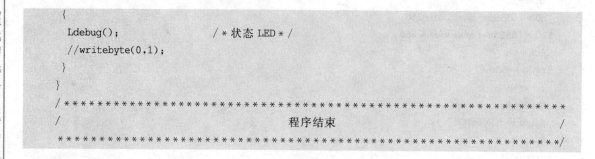

3.5 ADC

3.5.1 ADC 特性与结构

1. ADC 特性

nRF24LE1 的 ADC 具有最多 14 路输入通道（取决于封装形式），单端或差分输入，6 位、8 位、10 位或 12 位分辨率，单次转换模式的转换时间为 3 μs，连续模式可以采用 2 kbps、4 kbps、8 kbps 或 16 kbps 的采样速率。ADC 包含一个 1.2 V 的内部电压基准，也可采用外部基准或到 V_{DD} 的满量程基准；可以工作在单次采样模式或设定采样率的连续采样模式，测量电源电压模式；具有极低功耗，在采样速率为 2 kbps 时，电流仅为 0.1 mA。

2. ADC 内部结构

ADC 内部结构方框图如图 3.5.1 所示。

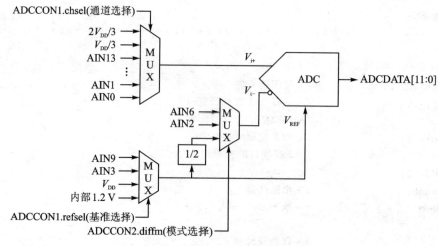

图 3.5.1　ADC 内部结构方框图

3.5.2 ADC 功能说明

1. 激活

当 pwrup 位被置位时,一个写操作到寄存器 ADCCON1 将自动启动一个转换。此时,如果 ADC 忙(已在操作中),未完成的转换将终止,而开始一个新的转换。写寄存器 ADCCON2 和 ADCCON3 将不会启动一个转换。不推荐在 ADC 忙时改变这些寄存器。

2. 输入选择

ADC 支持最多 14 路外部和 2 路内部输入通道,并可配置为单端或差分输入。输入通道由 chsel 位选择,通道 0~13(AIN0~AIN13)对应于 I/O 口的外部输入通道引脚端,通道 14 和 15 通道分别对应于内部 $V_{DD}/3$ 和 $2V_{DD}/3$。外部输入通道的数目取决于不同的封装。

配置 diffm 位可以选择单端或差分模式。在单端模式,输入电压范围从 0 V 到基准电压 V_{REF};在差分模式,输入电压范围为 $-V_{REF}/2$~$+V_{REF}/2$。在差分模式,AIN2 或 AIN6 可用来作为反相输入端,而使用 chsel 位选择同相输入端。共模电压范围必须在 V_{DD} 的 25%~75% 之间。

内部 $V_{DD}/3$ 和 $2V_{DD}/3$ 输入通道通常用来测量供电电压或校准偏移及增益误差。

3. 基准选择

全量程范围由 refsel 位控制,可以设置为内部基准(1.2 V)、外部基准或 V_{DD}。外部基准电压加到 AIN3 或 AIN9,但必须在 1.15~1.5 V 之间,具有片内高输入阻抗的 CMOS 缓冲区。

4. 分辨率

ADC 可以完成 6 位、8 位、10 位或 12 位转换,分辨率由 resol 位来选择。

5. 转换模式

cont 位用来选择单次或连续转换模式。在单次转换模式,ADC 完成一次转换后即停止;在连续转换模式,将按照已编程的采样率连续采样。

单次转换模式的时序图如图 3.5.2 所示,转换由写寄存器 ADCCON1 启动,忙标志位将在 4 个 16 MHz 时钟周期后置位 1,并在转换结果出现在寄存器 ADCDATH/ADCDATL 后清 0。转换结束后,将同时产生一个中断 ADCIRQ 到 MCU。

在默认情况下,ADC 单元将在转换完成后立刻进入掉电模式;也可以配置成转换结束后进入待机模式;还可以在一个可编程的时延后进入掉电模式。这样,如果一个新的转换在可编程的时延内发生,则可以减少启动时间。配置 rate 位可以选择工作性能。

注意:自动掉电模式将不会清除 pwrup 位,所选择的输入口继续配置为模拟输入,直到 pwrup 位由软件清除。

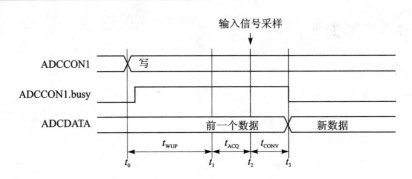

图 3.5.2 单次转换模式的时序图

一个 A/D 转换分为唤醒、信号采样和转换 3 个过程。唤醒时间取决于 ADC 启动前是处于掉电模式还是待机模式。如果处于掉电模式,则唤醒时间 $t_{WUP}=15\ \mu s$;如果处于待机模式,则唤醒时间为 $t_{WUP}=0.6\ \mu s$。

在唤醒过程结束时($t=t_1$),采样电容切换到模拟输入引脚端进行采样,并在整个采样阶段保持连接。在采样过程结束时($t=t_2$),信号被采样,这个过程的时间 $t_{ACQ}=0.75\ \mu s$、$3\ \mu s$、$12\ \mu s$、$36\ \mu s$,由 t_{ACQ} 位选择。

最后的过程是将模拟采样值转换为对应的 N 位数字。转换时间取决于所选择的分辨率:$t_{CONV}=1.7\ \mu s$、$1.9\ \mu s$、$2.1\ \mu s$ 和 $2.3\ \mu s$,分别对应于 6 位、8 位、10 位和 12 位转换。

连续转换模式的时序图如图 3.5.3 所示。除了新转换将按照所设置的速率连续进行外,在连续转换模式与单次转换模式下是相同的。在两次转换之间,A/D 转换器将进入掉电模式,以降低功耗。采样率由 rate 位设定为 2 ksps、4 ksps、8 ksps 或 16 ksps。

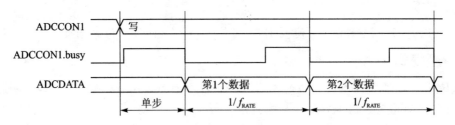

图 3.5.3 连续转换模式的时序图

6. 输出数据编码

对于单端输入方式的转换,ADC 采用直接二进制编码。输入电压≤0 V,对应为全 0 码 000…00;输入电压≥V_{REF} 基准电压,对应为全 1 码 111…11;中间值对应为 100…00。

对于差分输入方式的转换,ADC 采用带偏移的二进制编码。差分输入电压≤$-V_{REF}/2$,对应为全 0 码 000…00;输入电压≥$+V_{REF}/2$,对应为 1 码 111…11;零电位电压对应于 100…00。

ADCCON3 寄存器包含 3 个溢出位：当 ADC 欠量程时，uflow 置位；当 ADC 过量程时，oflow 置位；ADC 的量程范围是 uflow 与 oflow 的逻辑"或"。

7. 驱动模拟输入

每次当采样过程开始时，内部采样电容切换到输入引脚，模拟输入引脚端将会有一个小的电流瞬变，因此由内部电路来控制干扰，使输入在转换前趋于稳定是很重要的。除非输入由一个足够快的运算放大器所驱动，选择一个比最短采样时间更长的周期来确保稳定是非常必要的。应注意的是，这也相应地扩展了转换所需时间，延长了 ADC 返回掉电模式所需的时间。如果电流消耗是一个重要的问题，则应尽可能缩短采样时间。

推荐的采样时间和源阻抗以及电容的关系如图 3.5.4 所示。例如，一个无源的信号源和 10 位转换，其源阻抗为 100 kΩ，模拟输入引脚端的片外电容为 10 pF，可以从图 3.5.4 中读出，推荐的采样时间为 12 μs。

作为选择，一个大的电容可以连接在模拟输入引脚端和 V_{SS} 之间，它将提供所有的电流给采样电容，在这种情况下，即使源阻抗很高，也可以使用最短采样时间。推荐使用 33 nF 或更大的电容。

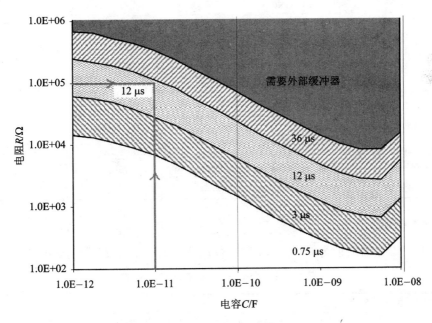

图 3.5.4　推荐的采样时间和源阻抗及电容的关系（10 位转换）

3.5.3　ADC 特殊功能寄存器

MCU 通过 5 个寄存器对 ADC 单元进行控制和访问：ADCCON1、ADCCON2、ADC-

CON3、ADCDATH 和 ADCDATL。ADCCON1、ADCCON2 和 ADCCON3 包含配置设置位和控制位。转换结果保存在寄存器 ADCDATH 和 ADCDATL 中。

1. ADCCON1

ADCCON1 寄存器地址为 0xD3,复位值为 0x00。寄存器各位功能如表 3.5.1 所列。

表 3.5.1 ADCCON1 寄存器各位功能

位	名 称	类 型	功 能
[7]	pwrup	R/W	加电控制。 0:ADC 掉电 1:加电到 ADC 和配置选择引脚端作为模拟输入
[6]	busy	R	ADC 忙(busy)标志。 0:没有转换在进行中。 1:转换在进行中。 当转换结果保存在 ADCDATH / ADCDATL 寄存器时,忙(busy)标志位清除
[5:2]	chsel	R/W	输入通道选择。 0000:AIN0。 0001:AIN1。 ⋮ 1101:AIN13。 1110:$V_{DD}/3$。 1111:$2V_{DD}/3$
[1:0]	refsel	R/W	电压基准选择。 00:内部 1.2 V 基准电压。 01:V_{DD}。 10:在 AIN3 的外部基准电压。 11:在 AIN9 的外部基准电压

2. ADCCON2

ADCCON2 寄存器地址为 0xD2。寄存器各位功能如表 3.5.2 所列。

表 3.5.2 ADCCON2 寄存器各位功能

位	名 称	类 型	功 能
[7:6]	diffm	R/W	选择单端或者差分模式。 00:单端模式。 01:差分模式,使用 AIN2 作为反相输入端。 10:差分模式,使用 AIN6 作为反相输入端。 11:未使用

续表 3.5.2

位	名称	类型	功能
[5]	cont	R/W	选择单次或者连续模式。 0：单次模式。 1：连续模式，采样速率由 rate 定义
[4:2]	rate	R/W	选择在连续模式的采样速率。 000：2 ksps。 001：4 ksps。 010：8 ksps。 011：16 ksps。 1xx：保留
[1:0]	tacq	R/W	输入采集窗口的时间(t_{ACQ})。 00：0.75 μs。 01：3 μs。 10：12 μs。 11：36 μs

3. ADCCON3

ADCCON3 寄存器地址为 0xD1，复位值为 0x00。寄存器各位功能如表 3.5.3 所列。

表 3.5.3 ADCCON3 寄存器各位功能

位	名称	类型	功能
[7:6]	resol	R/W	ADC 分辨率。 00：6 位。 01：8 位。 10：10 位。 11：12 位
[5]	rljust	R/W	选择左或右对齐在 ADCDATH / ADCDATL 中的数据。 0：左对齐数据。 1：右对齐数据
[4]	uflow	R	置位时，ADC 欠量程(转换结果全为 0)
[3]	oflow	R	置位时，ADC 过量程(转换结果全为 1)
[2]	range	R	置位时，ADC 过量程或者欠量程(等于 oflow 或 uflow)
[1:0]	—	—	未使用

4. ADCDATH/ADCDATL

ADCDATH 寄存器地址为 0xD4,复位值为 0x00。寄存器各位功能如表 3.5.4 所列。ADCDATL 寄存器地址为 0xD5,复位值为 0x00。寄存器各位功能如表 3.5.5 所列。ADC 左对齐或右对齐的输出数据如表 3.5.6 所列。

表 3.5.4 ADCDATH 寄存器各位功能

位	名称	类型	功能
[7:0]	—	R	ADCDATA 左对齐或右对齐的高字节

表 3.5.5 ADCDATL 寄存器各位功能

位	名称	类型	功能
[7:0]	—	R	ADCDATA 左对齐或右对齐的低字节

表 3.5.6 ADC 左对齐或右对齐的输出数据

rljust	resol	ADCDATH[7:0]	ADCDATL[7:0]
0	00	ADCDATA[5:0]	0
0	01	ADCDATA[7:0]	0
0	10	ADCDATA[9:0]	0
0	11	ADCDATA[11:0]	0
1	00	0	ADCDATA[5:0]
1	01	0	ADCDATA[7:0]
1	10	0	ADCDATA[9:0]
1	11	0	ADCDATA[11:0]

3.5.4 ADC 模拟电压输入电路

一个 nRF24LE1 ADC 模拟电压输入电路如图 3.5.5 所示,改变电位器 R_T 的位置,可以改变模拟输入端的输入电压。对于不同封装形式的 nRF24LE1 芯片,其 ADC 模拟输入端定义的引脚端数量不同。例如:对于 24 引脚 4 mm×4 mm 封装的芯片为 P0.0~P0.6(AIN0~AIN6),对于 32 引脚 5 mm×5 mm 封装的芯片为 P0.0~P0.7 和 P1.0~P1.2(AIN0~AIN10),对于 48 引脚 7 mm×7 mm 封装的芯片为 P0.0~P0.7 和 P1.0~P1.5(AIN0~AIN13)。

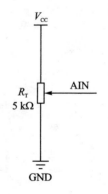

图 3.5.5 ADC 模拟电压输入电路

3.5.5 ADC 示例程序流程图

ADC 示例程序流程图如图 3.5.6 所示。

3.5.6 ADC 示例程序

P0.0 是 nRF24LE1 的 ADC 转换的模拟输入端 0，P0.1 是模拟输入端 1，P0.2 是模拟输入端 2。其中，P0.0 连接到 GND，P0.1 连接到图 3.5.5 的 AIN 引脚端，P0.2 连接到 V_{CC}。即 P0.0 的电压等于地的电压，P0.1 的电压可调，P0.2 的电压为 nRF24LE1 的电源电压，ADC 分别采样，输出结果是 ADC 通道 0 的采样值最小，通道 2 的采样值最大，通道 1 的采样值介于中间。

ADC 示例程序源代码如下：

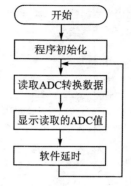

图 3.5.6 ADC 示例程序流程图

```
/***************************************************
/头文件包含
***************************************************/
#include "reg24le1.h"
#include "conf.h"
/***************************************************
/宏定义
***************************************************/
#define BIT0    0x01
#define BIT1    0x02
#define BIT2    0x04
#define BIT3    0x08
#define BIT4    0x10
#define BIT5    0x20
#define BIT6    0x40
#define BIT7    0x80
#define TOASC(dat) (dat + '0')         /*将数字转成对应的 ASCII 码*/
/***************************************************
/函数名称：adc_config()
/函数功能：nRF24LE1 的 ADC 功能初始化
/输入参数：无
/返回参数：无
***************************************************/
void adc_config()
{
    ADCCON2 = 0x00;                    /*设置成单步转换并使能,速度为 2 ksps*/
    ADCCON3 |= 0xE0;                   /*精度为 12 位,数据右对齐*/
    ADCDATH &= 0xF0;                   /*转换结果寄存器清 0*/
    ADCDATL &= 0x00;
```

第3章 nRF24LE1 的接口与应用

```
    P0DIR |= 0x07;                              /* 设置 ADC 转换的输入通道为 0、1、2 */
    P0 &= 0xF8;                                 /* 端口初始化为低电平 */
}

/****************************************************************
/函数名称：uart_init()
/函数功能：nRF24LE1 异步串口初始化
/输入参数：无
/返回参数：无
****************************************************************/
void uart_init()
{
    CLKCTRL = 0x28;                             /* 使用 XCOSC 16 MHz 时钟源 */
    CLKLFCTRL = 0x01;

    P0DIR &= 0xF7;                              /* 配置 P0.3(TXD)为输出 */
    P0DIR |= 0x10;                              /* 配置 P0.4(RXD)为输入 */
    P0 |= 0x18;
    S0CON = 0x50;
    PCON |= 0x80;                               /* 设置波特率倍增位 */
    WDCON |= 0x80;                              /* 选用内部波特率发生器 */

    S0RELL = 0xFB;
    S0RELL = 0xF3;                              /* 设置波特率为 38 400 */
}
/****************************************************************
/函数名称：send()
/函数功能：通过串口发送一个字符
/输入参数：ch 为待发送的字符
/返回参数：无
****************************************************************/
void send(char ch)                              /* 通过串口发送一个字符 */
{
    S0BUF = ch;
    while(!TI0);                                /* 等待发送完成 */
    TI0 = 0;
}
/****************************************************************
/函数名称：readadc()
/函数功能：读取指定通道的 ADC 转换结果
/输入参数：pip_num 为欲读取 ADC 转换的通道
/返回参数：选定通道上 ADC 转换得到的结果
```

```c
******************************************************/
unsigned int readadc(unsigned char * pip_num)
{
 unsigned int res = 0;
 static unsigned char num = 0;
 ADCCON1 = BIT7 + (num << 2) + BIT0;      /* 设置转换的通道,设置参考以及启动 */
 while(!(ADCCON1&BIT6));                   /* 等待启动转换 */
 while((ADCCON1&BIT6));                    /* 等待完成转换 */
 res = ADCDATL + (ADCDATH&0x0F) * 256;     /* 读取 ADC 转换的结果 */
 * pip_num = num;                          /* 保存当前 ADC 转换的通道号 */
 num ++ ;
 num = num % 3;                            /* 计算下一个需要转换的通道的通道号 */
 return res;
}


/**************************************************************
/函数名称: delay()
/函数功能: 软件延时
/输入参数: dx 为软件延时的时间数
/返回参数: 无
******************************************************/
void delay(unsigned int dx)
{
unsigned int di;
  for(;dx>0;dx--)
    for(di = 120;di>0;di--)
        {
          ;
        }
}
/**************************************************************
    /主函数
******************************************************/
void main(void)
{
uart_init();                              /* 串口初始化 */
adc_config();                             /* nRF24LE1 的 ADC 转换的初始化 */
while(1)
   {
    delay(20000);
    send('\n');
    result = readadc(&ch_num);            /* 函数读取通道的 ADC 转换结果,把通
```

```
            send(TOASC(ch_num));                    道号存在 ch_num 中,结果返回在 result 中 */
            send(' ');                              /*显示通道号*/
            send(' ');                              /*显示空格*/
            send(TOASC(result/1000));               /*显示对应通道的转换结果的千位*/
            send(TOASC((result%1000)/100));
            send(TOASC((result%100)/10));
            send(TOASC(result%10));
            send('\n');
        }                                           /*换行*/
    }
/***************************************************************
/                        程序到此结束                            /
****************************************************************/
```

程序被编译下载到 nRF24LE1 里,复位 nRF24LE1。连接计算机的串口和 nRF24LE1 的串口,串口波特率设置为 38 400,可以在计算机的显示器上看见格式为"* ****"的输出内容,而且在不断地更新。前面一个"*"的数字代表当前 ADC 转换通道号,后面"****"的数字代表该通道上 ADC 的采样值,如图 3.5.7 所示。

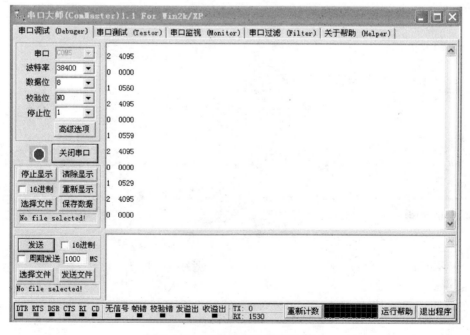

图 3.5.7　ADC 转换结果串口显示

3.6 模拟比较器

3.6.1 模拟比较器特性与结构

模拟比较器可用作唤醒源。允许通过引脚端,以差分或单端模拟输入来触发并且唤醒系统,轨到轨的输入电压范围,单端方式的阈值电平可编程为 V_{DD} 或来自引脚端的基准电压的 25%、50%、75% 或 100%。输出极性可编程。比较器具有非常低的功耗,电流消耗为 0.75 μA,可以工作在寄存器维持模式和存储器维持/定时器开启模式。

模拟比较器内部结构方框图如图 3.6.1 所示。

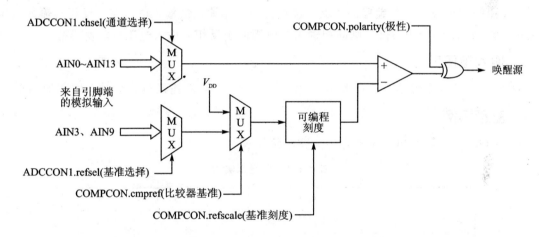

图 3.6.1 模拟比较器内部结构方框图

3.6.2 模拟比较器功能

1. 激 活

置位寄存器 COMPCON 的使能位将使能比较器。在寄存器维持模式和存储器维持/定时器开启模式,比较器是有效(激活)的,比较器不能工作在其他模式。使用比较器时,32 kHz 时钟源必须保持有效(激活)。

2. 输入选择

根据不同的封装,最多有 14 个不同的输入引脚端可以作为比较器的同相输入端,配置寄存器 ADCCON1 的 chsel 位来选择 AIN0~AIN13 作为输入。

注意:1110 和 1111 是禁用的值,如果使用了该值,则比较器的同相输入端将是浮置的。

ADCCON1 的 pwrup 位无须设置。

3. 基准源选择

比较器的反相输入端可以连接到 V_{DD} 或来自引脚端的基准电压(来自 AIN3 或 AIN9)的 25%、50%、75% 或 100%。配置寄存器 COMPCON 的 refscale 位可以选择比例系数。若使用 V_{DD} 作为基准源,则设置 cmpref 为 0;若使用外部基准源,则设置 cmpref 为 1。配置 ADCCON1 寄存器的 refsel 位,选择外部基准输入是 AIN3 或 AIN9。

注意:00 和 01 是 refsel 的禁用值,如果使用该值,将会导致比较器的反相输入引脚浮置。差分输入模式设置 refscale 为 100%,并选择 AIN3 或 AIN9 作为反相输入。

4. 输出极性

比较器的输出极性是可编程的。默认设置是当同相输入端输入电压高于反相端输入电压时触发唤醒;如果 polarity 置位,则同相输入端电压低于反相端输入电压时触发唤醒。

5. 输入电压范围

AIN0~AIN13 端的输入电压范围为 V_{SS}~V_{DD}+100 mV,输入电压绝对不能高于 3.6 V。

6. 配置示例

ADCCON1 和 COMPCON 寄存器配置示例如表 3.6.1 所列。

表 3.6.1 配置示例

唤醒标准	ADCCON1		COMPCON		
	chsel	refsel	polarity	refscale	cmpref
AIN0>0.25V_{DD}	0000	xx	0	00	0
AIN13<0.5V_{DD}	1101	xx	1	01	0
AIN2>0.75AIN3	0010	10	0	10	1
AIN3<AIN9	0011	11	1	11	1

7. 驱动模拟输入

比较器有一个受 32 kHz 时钟频率控制的开关电容,推荐在模拟输入引脚与 V_{SS} 之间连接 330 pF 的旁路电容,这将减少开关时导致的电压瞬变。当信号源的输出阻抗大于 100 kΩ 时,该电容对信号的影响可以被忽略。比较器的输入偏置电流低于 100 nA。

3.6.3 模拟比较器特殊功能寄存器

模拟比较器通过两个寄存器来控制。ADCCON1 配置外部引脚的多路复用,其他功能由

COMPCON 寄存器控制。

COMPCON 寄存器地址为 0xDB，复位值为 0x00，各位功能如表 3.6.2 所列。

表 3.6.2 COMPCON 寄存器各位功能

位	名称	类型	功能
[7:5]	—	—	未使用
[4]	polarity	R/W	输出极性。 0：同相。 1：反相
[3:2]	refscale	R/W	基准电压比例。 00：25%。 01：50%。 10：75%。 11：100%
[1]	cmpref	R/W	基准电压选择。 0：V_{DD}。 1：外部基准，来自 AIN3 或 AIN9
[0]	enable	R/W	比较器使能控制。 0：不使能比较器。 1：使能比较器和配置所选择的引脚端作为模拟输入

3.6.4 模拟比较器示例程序流程图

模拟比较器是一个能将 nRF24LE1 从低功耗模式唤醒的一个唤醒源。模拟比较器通过外部引脚来唤醒 nRF24LE1 进入到激活状态。

本示例程序流程图如图 3.6.2 所示。

3.6.5 模拟比较器示例程序

本示例程序首先使 nRF24LE1 在启动后定时 10 s，进入低功耗模式，等待模拟输入端输入高电平，将其唤醒，恢复正常的工作。

示例程序的源代码如下：

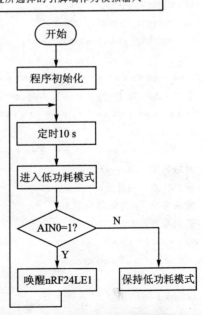

图 3.6.2 模拟比较器示例程序流程图

第3章 nRF24LE1 的接口与应用

```c
/****************************************************************/
#include "reg24le1.h"
/****************************************************************
/宏定义 nRF24LE1 各种低功耗模式
*****************************************************************/
#define LED           P10
#define ONESTEP       25000
#define deepsleep     0x01
#define timeroff      0x02
#define timeron       0x03
#define retention     0x04
#define standby       0x07
/****************************************************************
/函数原型声明
*****************************************************************/
void compcof();
void lightled();
void io_config();
void delay(unsigned char x);
void mcufallasleep(char which);
void puts( char *  s);
void t0_init();
void getwakeuptype();
void putchar( unsigned char dat);
void uart_init();
/****************************************************************
/定义 sleep 标志,当 sleep 为 1 时,nRF24LE1 就会进入低功耗模式
*****************************************************************/
int sleep = 0;
/****************************************************************
/函数名称:delay()
/函数功能:软件延时
/输入参数:x 为软件延时数
/返回参数:无
*****************************************************************/
void delay(unsigned char x)
{
unsigned char di;
    for(;x>0;x--)
```

```
    for(di = 120;di>0;di--)
    {
        ;
    }
}
/*****************************************************************
/函数名称: io_config()
/函数功能: 初始化 nRF24LE1 的 I/O 口
/输入参数: 无
/返回参数: 无
*****************************************************************/
void io_config()
{
    P1DIR &= 0xFE;              /* 设置 P1.0 为输出 */
    P10 = 0;

    P0DIR |= 0x01;
    P00 = 0;

}
/*****************************************************************
/函数名称: compcof()
/函数功能: 配置模拟较器的配置寄存器
/输入参数: 无
/返回参数: 无
*****************************************************************/
void compcof()
{
    ADCCON1 = (0x80|0x02);      /* 设置模拟输入通道和参考电压通道 */
    COMPCON |= 0x05;            /* 唤醒的电压高于 $0.75V_{DD}$ */
}
/*****************************************************************
/函数名称: lightled()
/函数功能: 闪烁 LED 指示灯
/输入参数: 无
/返回参数: 无
*****************************************************************/
void lightled()
{
```

```
    LED = !LED;
    delay(5000);
}
/*****************************************************************
/函数名称: uart_init()
/函数功能: nRF24LE1 异步串口初始化
/输入参数: 无
/返回参数: 无
*****************************************************************/
void uart_init()
{
    EA = 0;                /* 关闭所有中断 */
    CLKCTRL = 0x28;        /* nRF24LE1 的 16 MHz 时钟源的设置 */
    CLKLFCTRL = 0x01;      /* nRF24LE1 的实时时钟设置 */
    P0DIR &= 0xF7;         /* 配置 P0.3(TXD)为输出 */
    P0DIR |= 0x10;         /* 配置 P0.4(RXD)为输入 */
    S0CON = 0x50;
    PCON |= 0x80;          /* 配置波特率倍增 */
    WDCON |= 0x80;         /* 设置为内部波特率发生器 */
    S0RELL = 0xF3;         /* 波特率设置为 38 400 */
    S0RELH = 0x03;
}
/*****************************************************************
/函数名称: t0_init()
/函数功能: 初始化定时器 0
/输入参数: 无
/返回参数: 无
*****************************************************************/
void t0_init()
{
    TMOD = 0x01;
    TH0 = (65536 - ONESTEP)/256;
    TL0 = (65536 - ONESTEP)%256;
    ET0 = 1;
    TR0 = 1;
}
/*****************************************************************
/函数名称: t0_service()
/函数功能: 处理定时器 0 中断
```

/输入参数：无
/返回参数：无
**/
```c
void to_service() interrupt 1
{
  static char num = 0, s = 0;
  TR0 = 0;
  TH0 = (65536 - ONESTEP)/256;      /*重置起始计数*/
  TL0 = (65536 - ONESTEP) % 256;
  num ++ ;
  if(num == 20)
  {
  num = 0;
  s ++ ;
  LED = !LED;
  puts("time left : ");
  putchar((10 - s + '0'));          /*显示剩余时间*/
  putchar('\n');
  if(s == 10)
  {sleep = 1;                        /*时间到,设置进入低功耗的标志位*/
   s = 0;
  }
  }

  if(sleep)
  {
  sleep = 0;
  mcufallasleep(timeron);
  }
  TR0 = 1;
}
```
/**
/函数名称：putchar()
/函数功能：串口发送一个字符
/输入参数：dat 为串口发送的字符
/返回参数：无
**/
```c
void putchar( unsigned char dat)
```

```c
{
    S0BUF = dat;
    while(!TI0);
    TI0 = 0;
}
/************************************************************
/函数名称：puts()
/函数功能：串口发送一个字符串
/输入参数：s为指向字符串的指针
/返回参数：无
************************************************************/
void puts ( char * s)
{
    while( * s != '\0')
    {
    putchar( * s);
    s ++ ;
    }
}
/************************************************************
/函数名称：mcufallasleep()
/函数功能：设置 nRF24LE1 的 MCU 的低功耗模式
/输入参数：mode 为设置的低功耗模式
/返回参数：无
************************************************************/
void mcufallasleep (char mode)
{
OPMCON = 0x02;
PWRDWN & = 0xF8;
PWRDWN |= mode;
}
/************************************************************
/函数名称：getwakeuptype()
/函数功能：获取并串口显示上一次低功耗模式
/输入参数：无
/返回参数：无
************************************************************/
void getwakeuptype()
{
```

```c
char type = 0;
type = PWRDWN&0x07;              /*读取低功耗模式的寄存器*/
switch(type)                     /*判断低功耗模式类型*/
{
case 0x00:puts("power off            \n");break;
case 0x01:puts("deep sleep           \n");break;
case 0x02:puts("Memory retention, timer off    \n");break;
case 0x03:puts("Memory retention, timer on     \n");break;
case 0x04:puts("Register retention   \n");break;
case 0x07:puts("standby              \n");break;
}
putchar('\n');                   /*换行*/
}
/*************************************************************
/主函数部分
*************************************************************/
void main(void)
{
uart_init();                     /*串口初始化函数*/
compcof();                       /*模拟比较器初始化函数*/
t0_init();                       /*定时器0初始化*/
io_config();                     /*nRF24LE1 的 I/O 口初始化*/
delay(10);
putchar('\n');
getwakeuptype();
putchar('\n');
puts("System starting,this is just a test of nRF24LE1 Analog comparator");
putchar('\n');
EA = 1;
while(1)
{
lightled();
}
}
```

程序编译完成后下载到 nRF24LE1 上运行，通过串口线连接 nRF24LE1 和计算机的串口，通过计算机上的串口调试软件，可以监控 nRF24LE1 的工作状态，如图 3.6.3 所示。nRF24LE1 启动后倒计时 10 s，计时完成之后，就进入低功耗模式，在这种低功耗模式下，通过给 P0.0 一个高电平，就可以把 nRF24LE1 唤醒，使之恢复正常的工作状态。

第 3 章 nRF24LE1 的接口与应用

图 3.6.3 显示器上显示的程序运行结果

3.7 PWM

3.7.1 PWM 结构与功能

nRF24LE1 包含有两个 PWM(Pulse-Width Modulation，脉冲宽度调制)通道，PWM0 和 PWM1 两个通道共享一个频率和分辨率寄存器，而占空比控制是独立的，每个通道各有单独的输出引脚 PWM0 和 PWM1。当 MCU 时钟频率为 16 MHz 时，输出频率范围为 4～254 kHz。

PWM 内部结构方框图如图 3.7.1 所示。

3.7.2 PWM 特殊功能寄存器

nRF24LE1 的 PWM 采用 3 个寄存器来控制 PWM 的两个通道。

PWMCON(PWM 控制寄存器)使能 PWM 功能和设置 PWM 周期长度(即在一个 PWM 周期内的脉冲数)。PWMCON 寄存器地址为 0xB2，复位值为 0，功能如表 3.7.1 所列。

第 3 章　nRF24LE1 的接口与应用

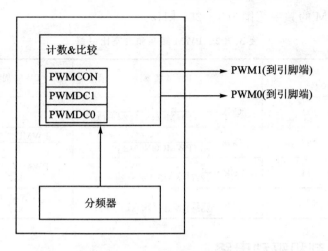

图 3.7.1　PWM 内部结构方框图

表 3.7.1　PWMCON 寄存器功能

位　数	类　型	复位值	功　能
[7:6]	R/W	00	使能/周期长度选择。 00：周期长度是 5 位。 01：周期长度是 6 位。 10：周期长度是 7 位。 11：周期长度是 8 位。
[5:2]	R/W	0000	PWM 频率分频系数选择。 (参考表 3.7.2)
[1]	R/W	0	选择 PWM1 输出通道。 0：PWM1 不使能。 1：PWM1 使能并有效。
[0]	R/W	0	选择 PWM0 输出通道。 0：PWM0 不使能。 1：PWM0 使能并有效。

寄存器 PWMDC0 和 PWMDC1 控制每个 PWM 通道的占空比。当其中一个寄存器被写入新值时，相对应的 PWM 信号将立刻改变为新的数值。这可以导致在一个 PWM 周期内有 4 次转变，但是在转变周期内，将会存在一个在旧采样值和新采样值间的直流电平。PWMDC0 用来控制 PWM 占空比，地址为 0xA1，8 位，复位值为 0。PWMDC1 用来控制 PWM 占空比，地址为 0xA2，8 位，复位值为 0。

通过 PWM 特殊功能寄存器来控制 PWM 的频率(或周期长度)和 PWM 占空比的示例如

表3.7.2所列。PWM的频率范围为4～254 kHz。

表3.7.2 PWM频率和占空比设置

PWMCON[7:6]（位数）	PWM频率	PWM周期
00 (5)	$C_{clk} \cdot \dfrac{1}{31(\text{PWMCON}[5:2]+1)}$	$\dfrac{\text{PWMDC}[4:0]}{31}$
01 (6)	$C_{clk} \cdot \dfrac{1}{63(\text{PWMCON}[5:2]+1)}$	$\dfrac{\text{PWMDC}[5:0]}{63}$
10 (7)	$C_{clk} \cdot \dfrac{1}{127(\text{PWMCON}[5:2]+1)}$	$\dfrac{\text{PWMDC}[6:0]}{127}$
11 (8)	$C_{clk} \cdot \dfrac{1}{255(\text{PWMCON}[5:2]+1)}$	$\dfrac{\text{PWMDC}}{255}$

3.7.3 电机控制和驱动电路

nRF24LE1具有两路PWM输出,当nRF24LE1工作在时钟频率16 MHz时,PWM输出的频率可以通过编程来改变。本示例中的按键和电机驱动电路如图3.7.2和图3.7.3所示。本示例程序可以根据按键输入开关的输入状态,改变PWM输出的占空比,从而实现电机转速的调节。

在图3.7.2中,V_{CC5V}表示的是5 V直流电压,$V_{CC7.2V}$表示的是7.2 V直流电压(注意,MC33886可以输入更高的电源电压)。OUT1和OUT0连接到直流电机。在电路PCB设计时,考虑电机运行过程中通过的电流较大。$V_{CC7.2V}$和GND连接有电源退耦电容器C_1和C_2。

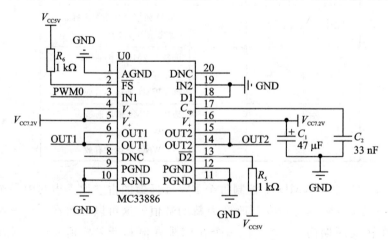

图3.7.2 MC33886电机驱动电路

nRF24LE1与外设的连接关系如表3.7.3所列。

第 3 章　nRF24LE1 的接口与应用

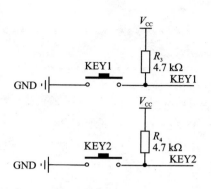

图 3.7.3　按键输入电路

表 3.7.3　nRF24LE1 与外设的连接关系

nRF24LE1 的 I/O 引脚	外设的网络
P0.2	PWM0
P1.0	KEY1
P1.1	KEY2

3.7.4　PWM 示例程序流程图

PWM 示例程序的流程图如图 3.7.4 所示。

本示例程序能够实现通过按键 KEY1 和 KEY2 调节 nRF24LE1 的 PWM 输出的占空比，从而达到控制直流电机的转速的目的。

MC33886 电机驱动芯片的 OUT1 和 OUT2 输出端的输出电压差的大小取决于 IN1 和 IN2 两个端口输入的 PWM 波的占空比。

在本示例程序中，通过按键 KEY1 和 KEY2 来改变 nRF24LE1 输出的 PWM 波的占空比。如果按下按键 KEY1，则 PWM 输出的占空比会降低；如果按下按键 KEY2，则 PWM 输出的占空比会升高。

在电机驱动电路中，IN1 连接到 PWM0 引脚端，IN2 连接到地（GND）。只要通过调节 PWM0 端口 PWM 波输出的占空比，即可实现直流电机的调速。

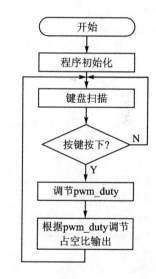

图 3.7.4　PWM 示例程序流程图

3.7.5　PWM 示例程序

PWM 示例程序的源代码如下：

```c
#include "pwm.h"
/*****************************************************************
/宏定义
*****************************************************************/
#define max 0xFF          /*定义最大的占空比允许值*/
#define min 0x00          /*定义最小的占空比允许值*/
/*****************************************************************
/函数名称: delay()
/函数功能: 软件延时
/输入参数: x为延时的时间数
/返回参数: 无
*****************************************************************/
void delay(unsigned int x)
{
unsigned char di;
   for(;x>0;x--)
    for(di=120;di>0;di--)
       {
         ;
       }
}
/*****************************************************************
/函数名称: keycheck()
/函数功能: 按键检测
/输入参数: 无
/返回参数: 按键键值
*****************************************************************/
unsigned char keycheck()
{
P1CON = 0xD0;             /*设置扫描 P1.0 引脚的电平*/
if(!P10)
{
delay(5);
if(!P10)
{
while(!P10);              /*等待按键释放*/
return FALSE;
}
}
```

```c
P1CON = 0xD1;                /*设置扫描 P1.1 引脚的电平*/
if(!P11)
{
delay(5);
if(!P11)
{
while(!P11);                 /*等待按键释放*/
return TRUE;
}
}
return NO_PRESS;             /*返回无按下状态*/
}
/***************************************************************
/函数名称：PWM_change()
/函数功能：用于设置输出 PWM 波的占空比
/输入参数：set_valnue 为占空比寄存器设置量
/返回参数：无
***************************************************************/
void PWM_change(unsigned int)
{
 PWMDC0 = PWMDC1 = set_valnue;
}


/***************************************************************
/函数名称：config()
/函数功能：初始化 nRF24LE1 的 PWM 功能
/输入参数：无
/返回参数：无
***************************************************************/
void config(void)
{
P1DIR |= 0x03;               /*设置 P1.0、P1.1 为输入引脚*/
P1 |= 0x03;                  /*设置初始化电平为高电平*/
P0DIR &= 0xF3;               /*设置 PWM0 和 PWM1 为输出*/
P02 = P03 = 0;               /*设置初始化电平为低电平*/
P1DIR &= 0xF3;
P12 = P13 = 0;
PWMCON = 0xC0;               /*关闭 PWM0 和 PWM1 的输出*/
PWMDC0 = 0x00;               /*设置 PWM0 占空比为 0*/
```

```c
    PWMDC1 = 0x00;              /*设置PWM1占空比为0*/
    PWMCON |= 0x03;             /*启动PWM0和PWM1输出PWM波*/
}
/***************************************************************
/主函数部分
****************************************************************/
void main()
{
unsigned int pwm_duty = 0x80;   /*定义PWM波输出占空比控制量*/
unsigned char flag = 0;         /*定义变量用来接收按键键值*/
config();                       /*初始化PWM功能*/
while(1)
  {
 flag = NO_PRESS;               /*设置按键初始状态为没有按下*/
 flag = keycheck();             /*获取按键*/
 if(flag == NO_PRESS)
 continue;                      /*如果没有按键按下,就继续扫描按键*/
 if(flag == TRUE)               /*如果返回为真,则增大占空比*/
  pwm_duty += 5;
  else if(flag == FALSE)        /*降低占空比*/
  pwm_duty -= 5;
  P12 = flag;
  P13 = flag;
  if(pwm_duty>max)              /*限制PWM的输出的最高占空比*/
  pwm_duty = max;
  if(pwm_duty<min)
  pwm_duty = min;
  PWM_change(pwm_duty);
   }
}
/***************************************************************
/                         程序结束                              /
****************************************************************/
```

程序运行后,电机具有一个初始的速度,P1.2和P1.3的输出电平最初值为低电平。如果连续地按按键KEY2,那么电机的转速会不断增加,直到达到当前条件下能够获得的最高转速为止。此时,如果按下按键KEY1,则P1.2和P1.3将输出低电平,电机的转速开始降低,直到输出的PWM的占空比为0为止。

第 4 章

nRF24LE1 的射频收发器与应用

4.1 nRF24LE1 的射频收发器

4.1.1 射频收发器内核结构与功能

nRF24LE1 使用与 nRF24L01+相同的 2.4 GHz GFSK 射频(RF)收发器内核,其内部结构方框图如图 4.1.1 所示,具有嵌入式协议引擎(增强型 ShockBurst)。射频收发器内核工作在 ISM 频段(2.400~2.483 5 GHz),适合各种超低功耗无线应用领域。

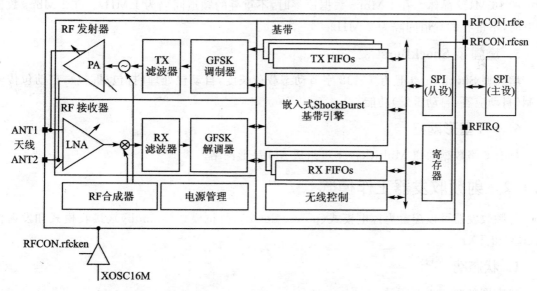

图 4.1.1　nRF24LE1 射频收发器内核结构方框图

通过对射频收发器的寄存器设置可以完成对射频收发内核的配置和操作。寄存器由 MCU 通过一个片上专用的 SPI 接口来访问,并可在各种 RF 收发器的所有模式下工作。

嵌入式协议引擎使能数据包通信,并且以各种灵活的模式进行,包括手动或自动的模式。

第 4 章 nRF24LE1 的射频收发器与应用

数据 FIFO 确保射频收发器和 MCU 之间的数据流通畅。

射频收发器内核具有如下功能：

1. 基本功能

射频收发器内核工作在 2.4 GHz ISM 频段，射频收发共用一个天线接口，GFSK 调制，250 kbps、1 Mbps、2 Mbps 的空中数据速率。

2. 发射功能

射频收发器内核的输出发射功率可编程，可编程为 0 dBm、−6 dBm、−12 dBm 或 −18 dBm。在 0 dBm 输出功率时，电流消耗为 11.1 mA。

3. 接收功能

具有集成的频道滤波器，数据速率为 2 Mbps 时，电流为 13.3 mA。接收灵敏度在数据速率为 2 Mbps 时为 −82 dBm，在 1 Mbps 时为 −85 dBm，在 250 kbps 时为 −94 dBm。

4. 射频合成器

片上集成了一个完全的频率合成器，具有 1 MHz 可编程分辨率，可使用低成本 ±60×10^{-6} 的 16 MHz 晶体。在 1 Mbps 数据速率时，不重叠的频道间隔为 1 MHz。在 2 Mbps 数据速率时，不重叠的频道间隔为 2 MHz。

5. 增强型 ShockBurst

增强型 ShockBurst 具有 1～32 字节动态载荷长度，自动包处理（打包/拆包），自动包传输控制（自动应答，自动重发）功能。

6. 6∶1 星形网络

具有 6 路数据通道 MultiCeiver 可组成 6∶1 星形网络。

4.1.2 射频收发器工作模式

射频收发器可以配置为掉电模式（power down）、待机模式（standby）、接收模式和发射模式（RX 和 TX）。

1. 状态图

状态图如图 4.1.2 所示。

状态图说明了射频收发器的工作模式及其功能。在复位时序结束后进入掉电模式。当射频收发器进入掉电模式时，MCU 仍可通过 SPI 和 RFCON 寄存器的 RFCON 位对其进行控制。

在图 4.1.2 中，有三个突出的不同状态类型。

第 4 章　nRF24LE1 的射频收发器与应用

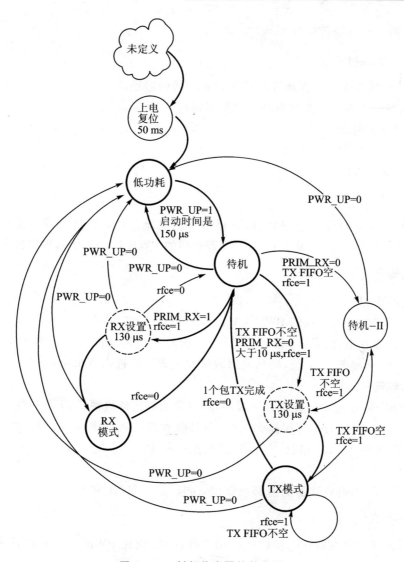

图 4.1.2　射频收发器的状态图

- 推荐的工作模式：建议该状态在正常操作时使用。
- 可能的工作模式：一种可能的运行状态，但不会在正常操作时使用。
- 过渡状态：在振荡器和 PLL 启动时，有时间限制的过渡状态。

在图 4.1.2 中：

⌇ 不规则的圆表示未定义。

◯ 粗线圆表示推荐的工作模式。

○ 细线圆表示可能的操作模式。

◌ 虚线圆表示过渡态。

➔ 粗线箭头表示在推荐的操作模式之间的路径。

→ 细线箭头表示在操作模式之间可能的路径。

CE=1：引脚信号条件。

PWR_DN=1：位状态的条件。

FIFO 空：系统信息。

2. 掉电模式

在掉电模式，射频收发器不使能，以获得最小的电流消耗。但所有寄存器的值将保持有效，而且 SPI 可以工作。通过设置 CONFIG 寄存器的 PWR_UP 位为 0，可以进入掉电模式。

3. 待机模式

(1) 待机模式-Ⅰ

通过设置 CONFIG 寄存器的 PWR_UP 位为 1，射频收发器将进入待机模式-Ⅰ。待机模式-Ⅰ可以使射频收发器在电流消耗比较小的情况下，有较短的启动时间。使能 rfce 位，进入正常工作模式。禁止(不使能)rfce 位，将从发射或接收模式直接进入待机模式-Ⅰ。

(2) 待机模式-Ⅱ

由于待机模式-Ⅱ需要额外的时钟缓冲器，电流消耗比待机模式-Ⅰ多一些。在 PTX 工作时，TX FIFO 为空，而且 rfce 位保持为 1 时，射频收发器进入待机模式-Ⅱ。如果一个新的数据包下载到 TX FIFO，PLL 将立刻启动，并在正常的 PLL 启动延时(130 μs)后开始发送数据包。

在待机模式下，所有寄存器的值都被保留，而且 SPI 可以正常工作。

4. 接收模式

接收模式是射频收发器作为接收机时的工作模式，设置 PWR_UP、PRIM_RX 和 rfce 位为 1 将进入接收模式。

在接收模式，接收机将解调无线信道所接收到的信号，而且将收到并解调的数据不断送到基带协议处理引擎，基带协议处理引擎将持续搜索有效的数据包。一旦发现有效的数据包(检测相匹配的地址和正确的 CRC 校验)，有效的数据载荷将出现在 RX FIFO 相应的存储单元中。如果 RX FIFO 已满，则所接收的数据将被丢弃。

射频收发器部分将一直处于接收机状态，直到 MCU 将其配置为待机模式-Ⅰ或掉电模式。但是如果使能了基带协议处理引擎中的自动协议功能(增强型 ShockBurst)，射频收发器将会进入其他模式来执行协议的功能。

在接收模式中,接收功率检测器(RPD)功能是可用的。当接收到的射频信号强度高于 -64 dBm 时,RPD 信号将置位为高电平。内部的 RPD 信号在送到 RPD 寄存器前会经过滤波。在 RPD 置位为高电平前,射频信号必须至少持续 40 μs。

5. 发射模式

发射模式用于发射数据包。设置 PWR UP 为 1,PRIM_RX 为 0,并且数据载荷已在 TX FIFO 中,同时 rfce 位出现一个时间大于 10 μs 的 1 状态,射频收发器进入发射模式。

射频收发器将保持在发射模式,直到完成一个数据包发送。如果 rfce=0,则射频收发器将进入待机模式-I。如果 rfce=1,则 TX FIFO 的状态将确定下一个操作。如果 TX FIFO 非空,则射频收发器将保持在发射模式并发射下一个包。当 TX FIFO 为空时,射频收发器将进入待机模式-II。在发射模式下,射频收发器的发射机 PLL 为开环状态。在发射模式下,发射时间一次不超过 4 ms,即使在增强型 ShockBurst 下也是如此。

射频收发器的接收/发射控制由射频收发器的 CONFIG 寄存器的 PRIM_RX 位来控制。

6. 工作模式配置

射频收发器的工作模式配置如表 4.1.1 所列。

表 4.1.1 射频收发器的工作模式配置

模 式	PWR_UP 寄存器	PRIM_RX 寄存器	rfce	FIFO 状态
接收模式	1	1	1	—
发射模式	1	0	1	数据在 TX FIFO。将清空 TX FIFO 的所有层
发射模式	1	0	—	数据在 TX FIFO。将清空 TX FIFO 的一层
待机模式-II	1	0	1	清空 TX FIFO
待机模式-I	1	—	0	没有正在进行传输的数据包
掉电模式	0	—	—	—

4.1.3 射频收发器空中速率

空中速率是指射频收发器在发射或接收时所使用的已调制的信号速率。对于 nRF24LE1,空中速率可设定为 250 kbps、1 Mbps 或 2 Mbps。使用较低的速率可以获得更好的接收灵敏度。但高速率可以获得更低的平均电流消耗,并减少在空中受到干扰和碰撞的机会。

空中速率由 RF_SETUP 寄存器中的 RF_DR 位来设定。相互通信的发射方和接收方必须设定为同一速率。射频收发器与 nRF24L01 完全兼容。为了与 nRF2401A、nRF2402、nRF24E1 及 nRF24E2 兼容,空中速率须设定为 250 kbps 或 1 Mbps。

4.1.4 射频收发器射频通道频率

射频通道的频率决定射频收发器所使用通道的中心频率,在速率为 250 kbps 或 1 Mbps 时,射频通道占用的带宽小于 1 MHz,而在速率为 2 Mbps 时,射频通道所占带宽小于 2 MHz。

射频收发器工作的频率范围为 2.400～2.525 GHz。可编程的射频通道频率分辨率为 1 MHz。

由于在 2 Mbps 通信速率时,占用带宽为超过通道频率分辨率,为了确保在 2 Mbps 速率下不出现重叠,通道频率间隔必须设定为 2 MHz 或者更宽一些。而在 1 Mbps 和 250 kbps 速率时,所占用带宽等于或低于射频通道频率分辨率。

射频通道频率由 RF_CH 寄存器设置,可以利用下面的公式计算得出:

$$f_0 = (2\,400 + \text{RF\_CH}) \text{MHz}$$

为确保相互通信,相互通信的发射方和接收方必须设定为同一频率。

4.1.5 接收功率检测

RPD(Received Power Detector,接收功率检测)位于寄存器 09 的位 0。在射频通道上接收到的功率电平高于 -64 dBm 时,RPD=1。如果接收到的信号功率电平低于 -64 dBm,则 RPD=0。

在接收模式,RPD 位可以在任何时候被读出。这使得可以随时监测当前射频通道上的接收功率电平。如果没有数据包接收,或 MCU 设定 rfce 位为 0,或者由 Enhanced ShockBurst 控制的接收时间超时,RPD 将被锁存。

当接收模式使能,并经过一个延时即 $T_{stby2a} + T_{delay\_AGC} = 130\,\mu s + 40\,\mu s$ 后,RPD 的状态将是正确的。由于接收的增益随着温度变化,因此 RPD 的检测电平阈值也会随温度变化,这个值在 $-40\ ℃$ 时减少约 5 dB,在 85 ℃ 时将增加 5 dB。

4.1.6 PA 控制

PA(Power Amplifier,功率放大器)控制用来设置射频发射的输出功率。在发射模式,PA 输出功率可编程为四级,如表 4.1.2 所列(条件:$V_{DD}=3.0$ V,$V_{SS}=0$ V,$T_A=27$ ℃,负载阻抗 $=15\,\Omega + j\,88\,\Omega$。)。PA 输出功率由 RF_SETUP 寄存器的 RF_PWR 位来控制。

表 4.1.2 PA 输出功率设置

RF_PWR	RF 输出功率/dBm	电流消耗/mA
11	0	11.1
10	-6	8.8
01	-12	7.3
00	-18	6.8

4.1.7 增强型 ShockBurst

增强型 ShockBurst 是一个以包为基础的数据链路层,其功能包括包的自动装配和设定装配时间、自动应答和自动重发。增强型 ShockBurst 能够完成超低功耗和高性能的通信,能有效改善双向和非双向系统无线通信的电源效率,而不需要主控制器进行复杂的操作。

nRF24LE1 的增强型 ShockBurst 的主要功能包括 1~32 字节的动态数据载荷、自动打包、自动进行数据包的相关处理、自动应答、自动重传及 1:6 星形网络的 6 数据通道 MultiCeiver 管理。

1. 增强型 Shockburst 数据包格式

增强型 ShockBurst 的数据包包括一个前置域(Preamble Field)、地址域(Address Field)、包控制域(Packet Control Field)、数据载荷域(Payload Field)和 CRC 校验域(CRC Field)。增强型 ShockBurst 数据包格式如图 4.1.3 所示,MSB(最高位)在左边。

前置域1字节	地址3~5字节	包控制域9位	数据载荷域0~32字节	CRC 1~2字节

图 4.1.3 增强型 ShockBurst 数据包格式

前置域是一个用于同步接收解调器的位序列。前置域为一字节长,内容为 01010101 或 10101010。如果地址位的第一位是 1,则前置域自动设为 10101010。如果地址位的第一位是 0,则前置域自动设为 01010101。这是为了确保有足够的过渡时间来稳定接收机操作。

地址域指的是接收机地址。地址用来确保数据包被相应的接收机正确地接收。地址可以通过 AW 寄存器配置为 3、4 或 5 字节长。

包控制域包括 6 位数据载荷长度(Payload Length)、2 位 PID(Packet Identity,包标识符)和 1 位 NO_ACK 标志(无应答标志)。9 位包控制域的格式如图 4.1.4 所示,MSB 在左边。

数据载荷长度6位	PID 2位	NO_ACK 1位

图 4.1.4 9 位包控制域的格式

6 位的数据载荷长度用来确定数据载荷的字节长度,数据载荷的字节长度可为 0~32 字节,如 000000=0 字节(仅使用一个空的 ACK 包),100000=32 字节,100001=忽略。数据载荷长度仅在动态载荷长度使能时使用。

2 位 PID 域用来确定接收到的数据包是新的还是重发的。PID 可以防止 PRX 端的 MCU 多次处理同一组数据载荷。在发射端,每通过 SPI 收到一个新的数据包,PID 就会加 1。PID 和 CRC 域用来确定一个接收到的包是新的还是重发的。当连续丢失几个包之后,PID 域可能变为相同的,所以一个包如果与上一个包有相同的 PID,将会比较这两个包的 CRC 校验是否相同,如果 CRC 校验也是相同的,将会被认为收到的是上一个包的拷贝(Copy)而被丢弃。

NO_ACK 位用来选择自动应答功能。该标志用来确定自动应答功能是否被使用。设定此位为 1,将告诉接收机无须自动应答。

数据载荷是用户所定义的包的内容,可以是 0～32 字节长度,加载到器件后即可发射出去。

增强型 ShockBurst 提供静态和动态两种处理载荷长度的方法。默认是静态数据载荷长度。静态数据载荷长度由接收端的 RX_PW_Px 寄存器确定,发射端的静态数据载荷长度由送入 TX_FIFO 的字节长度所确定,并且其长度必须与接收端 RX_PW_Px 寄存器设置的数值相一致。

DPL(Dynamic Payload Length,动态载荷长度)允许发射机发送不同长度的数据包到接收端,这就意味着系统可以处理不同的数据包,而不必衡量数据包的长度。

使用 DPL 功能,nRF24L01+可以自动解码所接收的包的长度,而不是采用 RX_PW_Px 寄存器的值。MCU 可以使用 R_RX_PL_WID 指令读取数据包的长度。

为了使能 DPL 功能,必须使能 FEATURE 寄存器的 EN_DPL 位。在接收模式,DYNPD 寄存器必须进行设置。当 PTX 使用 DPL 功能发送到 PRX 时,必须将 DYNPD 寄存器的 DPL_P0 位置位。

CRC(循环冗余校验)是数据包中使用的错误检测机制,它可以是 1 字节或 2 字节,用来计算地址、包控制域、数据载荷的校验和。

- 1 字节的 CRC 多项式为 x^8+x^2+x+1。初始值为 0xFF。
- 2 字节的 CRC 多项式为 $x^{16}+x^{12}+x^5+1$。初始值为 0xFFFF。

如果 CRC 校验失败,数据包将不会被增强型 ShockBurst 所接收。

2. PTX 操作

进入待机模式-Ⅰ后,增强型 ShockBurst 配置 PTX 的操作流程如图 4.1.5 所示。

3. PRX 操作

进入待机模式-Ⅰ后,增强型 ShockBurst 配置 PRX 的操作流程如图 4.1.6 所示。

4. MultiCeiver

MultiCeiver(多路接收)是接收模式下的一个特殊功能,如图 4.1.7 所示,可以连接 6 个并行的数据通道。在物理的 RF 通道中,一个数据通道称为逻辑通道。每个数据通道有其自己的物理地址(数据通道地址)由射频收发器进行解码。

射频收发器配置为 PRX(主接收)可以在一个 RF 频率通道下接收 6 路不同的数据通道,每个数据通道都有自己唯一的地址,并可以配置完成不同的功能。所有的数据通道都能够完成增强型 ShockBurst 功能。

第 4 章　nRF24LE1 的射频收发器与应用

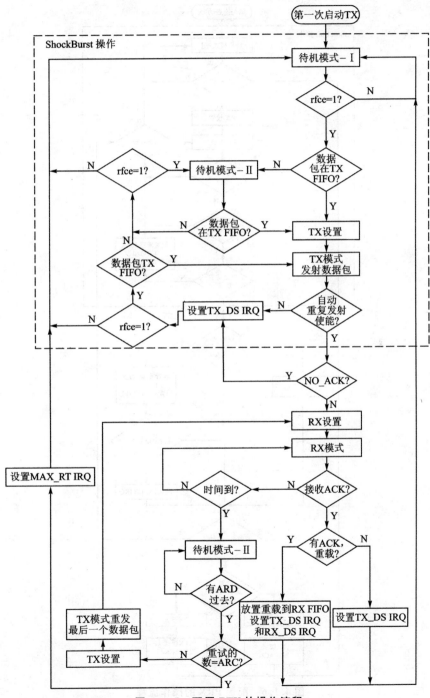

图 4.1.5　配置 PTX 的操作流程

第 4 章　nRF24LE1 的射频收发器与应用

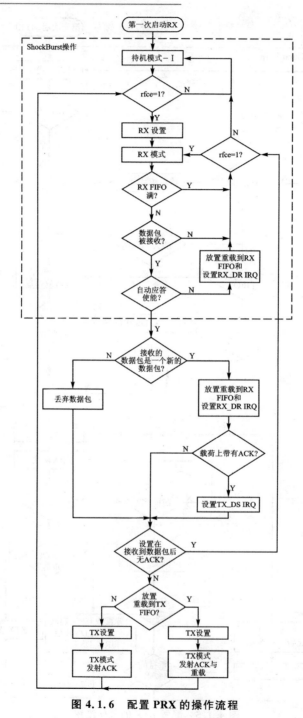

图 4.1.6　配置 PRX 的操作流程

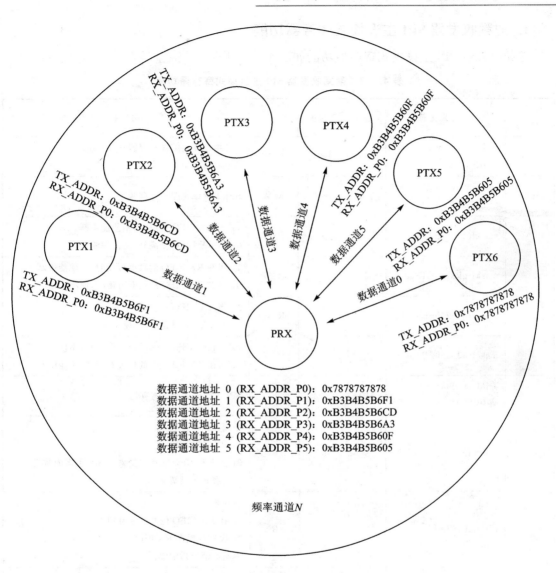

图 4.1.7 MultiCeiver 接收模式

有关增强型 ShockBurst 的更多内容请登录 www.nordicsemi.com，查询 nRF24LE1 Ultra-low Power Wireless System On-Chip Solution Preliminary Product Specification v1.6。

4.1.8 数据和控制接口

利用数据及控制接口可以访问射频收发器的所有功能。独立的 SFR（特殊功能寄存器）用来替代端口引脚。数据及控制接口与 nRF24L01＋芯片相同。

第 4 章 nRF24LE1 的射频收发器与应用

1. 射频收发器 SPI 主设模式寄存器功能

射频收发器 SPI 主设模式寄存器功能如表 4.1.3 所列。

表 4.1.3 射频收发器 SPI 主设模式寄存器功能

地 址 (Hex)	名称/助记符	位	复位值	类型	功 能
0xE4	spiMasterConfig0 SPIRCON0	[6:0]	0x01	R/W	SPI 主设模式配置寄存器 0 保留。不改变
0xE5	spiMasterConfig1 SPIRCON1	[3:0]	0x0F	R/W	SPI 主设模式配置寄存器 1
	maskIrqRxFifoFull	[3]	1	R/W	1：当 RX FIFO 是满时,不使能中断。 0：当 RX FIFO 是满时,使能中断
	maskIrqRxData-Ready	[2]	1	R/W	1：当在 RX FIFO 的数据可用时,不使能中断。 0：当在 RX FIFO 的数据可用时,使能中断
	maskIrqTxFifo-Empty	[1]	1	R/W	1：当 TX FIFO 是空时,不使能中断。 0：当 TX FIFO 是空时,使能中断
	maskIrqTxFifo-Ready	[0]	1	R/W	1：当在 TX FIFO 中的位置有用时,不使能中断。 0：当在 TX FIFO 中的位置有用时,使能中断
0xE6	spiMasterStatus SPIRSTAT	[3:0]	0x03	R	SPI 主设模式状态寄存器
	rxFifoFull	[3]	0	R	中断源。 1：RX FIFO 满。 0：在 RX FIFO 能够接受来自 SPI 更多的数据。 中断源消失后清除
	rxDataReady	[2]	0	R	中断源。 1：在 RX FIFO 位的数据可用。 0：没有数据在 RX FIFO。 中断源消失后清除
	txFifoEmpty	[1]	1	R	中断源。 1：TX FIFO 空。 0：数据在 TX FIFO 中。 中断源消失后清除
	txFifoReady	[0]	1	R	中断源。 1：在 TX FIFO 位置可用。 0：TX FIFO 满。 中断源消失后清除

续表 4.1.3

地址(Hex)	名称/助记符	位	复位值	类型	功 能
0xE7	spiMasterData SPIRDAT	[7:0]	0x00	R/W	SPI 主设模式数据寄存器。访问 TX（写）和 RX（读）FIFO 缓冲器，深度为 2 字节

射频收发器的 SPI 主设模式通过 SPIRCON1 寄存器配置。4 个不同的信号源可以产生中断，除非这些中断源已经在 SPIRCON1 中的相应位被屏蔽。SPIRSTAT 指示哪一个信号源所产生中断。SPIRSTAT 可以访问 TX（写）和 RX（读）FIFO 缓冲器，均为 2 字节深度。FIFO 是动态的，可以根据状态位的标志进行再装填，FIFO ready 表示 FIFO 可以接收数据；Data ready 表示 FIFO 可以提供至少 1 字节的数据。

2. RFCON 寄存器功能

RFCON 寄存器控制射频收发器的 SPI 从设模式片选信号（CSN）、片选信号（CE）和时钟使能信号 CKEN。其功能如表 4.1.4 所列。

表 4.1.4 RFCON 寄存器功能

地址	位	名 称	R/W	功 能
0xE8	[7:3]	—	—	保留
	[2]	rfcken	R/W	时钟使能（16 MHz）
	[1]	rfcsn	R/W	使能 RF 指令，0：使能
	[0]	rfce	R/W	使能射频收发器，1：使能

3. SPI 命令集

nRF24LE1 提供射频收发器 SPI 命令集。SPI 命令集请查询 nRF24LE1 Ultra-low Power Wireless System On-Chip Solution Preliminary Product Specification v1.6。每一个新命令必须从写 0 到 RFCON 寄存器的 rfcsn 位开始。SPI 命令通过写命令字到 SPIRDAT 寄存器来传送到射频收发器，第一个传输完成后，射频收发器可从 STATUS 寄存器读出来自 SPIRDAT 寄存器的内容。

串行移位的 SPI 命令顺序格式如下：

命令字——高位到低位（1字节）；

数据字节——低字节到高字节，每字节中高位先。

4. 数据 FIFO

数据 FIFO 存储发送的数据载荷（TX FIFO）或接收就绪的数据载荷（RX FIFO）。FIFO 在 PTX 和 PRX 模式下均可以访问。

在射频收发器中,TX 有三级,32 字节 FIFO;RX 有三级,32 字节 FIFO。

TX 和 RX 的 FIFO 有一个控制器,均可以通过 SPI 接口采用专用 SPI 命令来访问。TX FIFO 在 PRX 模式,可以存储为 3 个不同 PTX 操作的 ACK 包的载荷。在 1 个数据通道中,如果 TX FIFO 包含超过 1 个载荷,载荷的处理将遵循先进先出的原则。TX FIFO 在 PRX 模式,如果失去与其所对应的 PTX 的射频连接,载荷将被阻塞。在这种情况下,MCU 可以使用 FLUSH_TX 命令来清除刷新 TX FIFO。

RX FIFO 在 PRX 模式可以包含来自 3 个不同 PTX 操作的载荷,PTX 模式下的 TX FIFO 最多可以存储 3 个载荷。可使用如下的 3 个指令写 TX FIFO:

- W_TX_PAYLOAD;
- W_TX_PAYLOAD_NO_ACK(在 PTX 模式);
- W_ACK_PAYLOAD(在 PRX 模式)。

这 3 个指令可用来提供访问到 TX_PLD 寄存器。

在 PTX 和 PRX 模式,RX FIFO 能够利用 R_RX_PAYLOAD 指令读取,该指令提供访问到 RX_PLD 寄存器。当 MAX_RT IRQ 中断产生时,在 PTX 模式的 TX FIFO 的载荷不会被删除。

FIFO(RX 和 TX)结构方框图如图 4.1.8 所示。

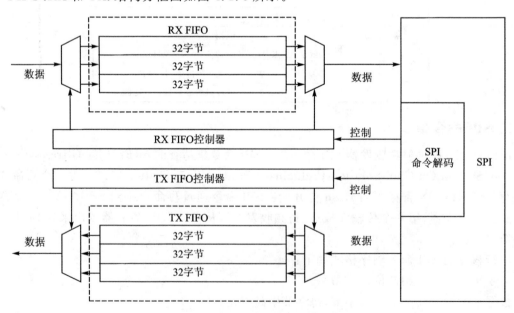

图 4.1.8　FIFO(RX 和 TX)结构方框图

可以读 FIFO_STATUS 寄存器来查看 TX FIFO 和 RX FIFO 是满或空。TX_REUSE (也在 FIFO_STATUS 寄存器中)可通过 SPI 指令 REUSE_TX_PL 来置位,通过 SPI 命令

W_TX_PAYLOAD 或 FLUSH TX 来复位。

5. 中 断

射频收发器可以发送中断给 MCU，当 TX_DS、RX_DR 或 MAX_RT 被 STATUS 寄存器中的状态机置高时，中断(RFIRQ)激活。当 MCU 写 1 到 STATUS 状态寄存器相应的中断源位时，RFIRQ 中断解除。CUNFIG 寄存器中的中断屏蔽用来选择可以激活 RFIRQ 的中断源，将相应中断源位置高将禁止该中断。所有中断源默认为使能。

注意：在 RFIRQ 由高电平到低电平过渡时，STATUS 寄存器中的 3 位数据通道信息被更新，所以在 RFIRQ 由高电平到低电平过渡时读取数据通道信息是不可靠的。

6. 寄存器图

用户可以通过 SPI 访问寄存器来配置和控制 nRF24LE1 的无线功能（使用读/写命令）。寄存器列表请查询 nRF24LE1 Ultra-low Power Wireless System On-Chip Solution Preliminary Product Specification v1.6。

注意：表中未定义的位是无用的，读出为 0。地址 18～1B 保留用作测试，改变它们将会使芯片发生故障。

4.2 射频收发器应用示例 1

4.2.1 无线传输结构形式

本示例使用两个 nRF24LE1 模块实现无线数据传输和数据接收，结构示意图如图 4.2.1 所示。计算机(PC 机)作为上位机，通过串口将需要发送的数据发给 nRF24LE1 发送端，nRF24LE1 发送端通过无线发射传输到 nRF24LE1 接收端，接收端将接收到的数据通过串口传送到与之连接的计算机上，从而实现一个无线的串口功能。

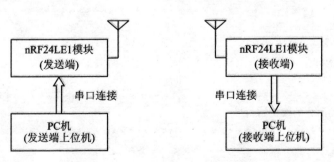

图 4.2.1 无线串口通信结构示意图

4.2.2 无线传输示例程序流程图

nRF24LE1 无线发送部分的程序流程图如图 4.2.2 所示,无线接收部分的程序流程图如图 4.2.3 所示。

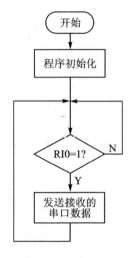
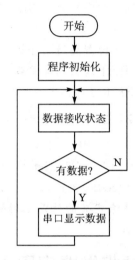

图 4.2.2　nRF24LE1 无线发送端程序流程图　　图 4.2.3　nRF24LE1 无线接收端程序流程图

4.2.3 无线传输示例程序

1. 初始化无线收发部分的数据结构

初始化无线收发部分的数据结构程序源代码如下:

```
/****************************************************************
/文件名称：wireless_api.h
****************************************************************/
#ifndef __WIRELESS_API_H__
#define __WIRELESS_API_H__
#include "config.h"
#include "reg24le1.h"
/****************************************************************
/初始化无线收发部分的数据结构
****************************************************************/
typedef struct
{
    unsigned char   nRecvAddr;
    unsigned char   nAutoAck;
```

```
        unsigned char    aAddr1[5];
        unsigned char    aAddr2[5];
        unsigned char    nPower;
        unsigned char    nChannel;
        unsigned char    nRetran;
        unsigned char    nLen;
        unsigned char    nOn;
        unsigned char    nART_Factor;
} SetupData;
/*****************************************************************
/部分函数的声明,通过这个头文件可以调用下面的函数
*****************************************************************/
unsigned char Rf24L01_Set_Init(SetupData * drc);
unsigned char Rf24L01_Polling_IRQ(unsigned char * rev_buf);
void wireless_init(void);
void Rf24L01_RxTx_Switch(unsigned char bMode);
static void Rf24L01_RX(unsigned char * prx_buf);
void Rf24L01_TX(unsigned char * ptx_buf,unsigned char nLen);
unsigned char Rf24L01_Set_Init(SetupData * drc);
#endif
/*****************************************************************
/                  头文件 wireless_api.h 到此结束                  /
*****************************************************************/
```

2. 无线收发模式的宏定义

无线收发模式的宏定义程序源代码如下:

```
/*****************************************************************
/文件名称:config.h
*****************************************************************/
#ifndef CONFIG_H__
#define CONFIG_H__
#include "reg24le1.h"
/*****************************************************************
/无线收发模式宏定义以及布尔常量定义
*****************************************************************/
#define  PTX      0x00
#define  PRX      0x01
#define  TRUE     0x01
#define  FALSE    0x00
```

```
/************************************************************
/包含函数原型,通过包含本头文件
*************************************************************/
void Timer_Ms(int nTime);
void Uart_Init(void);
void Timer_10Us(unsigned char nTime);
void hal_wdog_init(unsigned int start_value);
void hal_wdog_restart(void);
void putchar( unsigned char dat);
void puts( char * s);
char getch(void);
void system_init(void);
unsigned char keycheck(void);
#endif
/************************************************************
/                    头文件 config.h 到此结束                /
*************************************************************/
```

3. 宏定义一些变量和函数

宏定义一些变量和函数如下:

```
/************************************************************
/文件名称: rf24le1.h
/文件说明:宏定义一些变量和一些简单的函数,包含无线传输功能的函数原型,使这些函数被调用时
         更加灵活方便
*************************************************************/
#ifndef Rf_24L01H
#define Rf_24L01H

#include "config.h"
#include "reg24le1.h"
#include "string.h"

/************************************************************
/宏定义部分,定义一些特殊的函数
*************************************************************/
#define CSN_LOW()      RFCSN = 0;
#define CSN_HIGH()     RFCSN = 1;
#define CE_LOW()       RFCE = 0;Timer_10Us(1);
#define CE_HIGH()      RFCE = 1;
```

```c
#define CE_PULSE()    CE_HIGH();Timer_10Us(2);CE_LOW();
/****************************************************************
/ 基频定义
****************************************************************/
#define Def_Base_Channel    122
/****************************************************************/
/nRF24LE1 的无线收发部分电源标志
****************************************************************/
#define RF_POW_ON         0x02
#define RF_POW_OFF        0x00
/****************************************************************
/定义中断标志
****************************************************************/
#define IDLE              0x00        /*空闲*/
#define MAX_RT            0x10        /*重传次数最大*/
#define TX_DS             0x20        /*发送完成*/
#define RX_DR             0x40        /*接收完成*/
/****************************************************************
/宏定义无线传输中用到的指令
****************************************************************/
#define READ_REG          0x00        /*定义寄存器读指令*/
#define WRITE_REG         0x20        /*定义寄存器写指令*/
#define ACTIVATE          0x50        /*定义写指令*/
#define READ_PAYLAODLEN   0x60        /*定义读传输数据宽度指令*/
#define RD_RX_PLOAD       0x61        /*定义读接收负载寄存器地址*/
#define WR_TX_PLOAD       0xA0        /*定义写发送负载寄存地址*/
#define WR_TX_PLOAD_ACK   0xA8        /*写带应答的传输*/
#define WR_TX_PLOAD_NOACK 0xB0        /*特殊包关闭应答*/
#define FLUSH_TX          0xE1        /*定义刷新发送缓冲指令*/
#define FLUSH_RX          0xE2        /*定义刷新接收缓冲指令*/
#define REUSE_TX_PL       0xE3        /*定义复用发送寄存器指令*/
#define NOP               0xFF        /*定义空操作*/
/****************************************************************
/(SPI) nRF24LE1 与无线相关的寄存器定义
****************************************************************/
#define CONFIG            0x00        /*寄存器 CONFIG 的定义*/
#define EN_AA             0x01        /*寄存器 EN_AA 的定义*/
#define EN_RXADDR         0x02        /*寄存器 EN_RXADDR 的定义*/
#define SETUP_AW          0x03        /*寄存器 SETUP_AW 的定义*/
```

第4章 nRF24LE1 的射频收发器与应用

```
#define SETUP_RETR         0x04        /*寄存器 SETUP_RETR 的定义*/
#define RF_CH              0x05        /*寄存器 RF_CH 的定义*/
#define RF_SETUP           0x06        /*寄存器 RF setup 的定义*/
#define STATUS             0x07        /*寄存器 STATUS 的定义*/
#define OBSERVE_TX         0x08        /*寄存器 OBSERVE_TX 的定义*/
#define CD                 0x09        /*传输监测寄存器地址定义*/
#define RX_ADDR_P0         0x0A        /*接收通道0寄存器地址定义*/
#define RX_ADDR_P1         0x0B        /*接收通道1寄存器地址定义*/
#define RX_ADDR_P2         0x0C        /*接收通道2寄存器地址定义*/
#define RX_ADDR_P3         0x0D        /*接收通道3寄存器地址定义*/
#define RX_ADDR_P4         0x0E        /*接收通道4寄存器地址定义*/
#define RX_ADDR_P5         0x0F        /*接收通道5寄存器地址定义*/
#define TX_ADDR            0x10        /*发射寄存器地址定义*/
#define RX_PW_P0           0x11        /*接收通道0负载宽度寄存器地址定义*/
#define RX_PW_P1           0x12        /*接收通道1负载宽度寄存器地址定义*/
#define RX_PW_P2           0x13        /*接收通道2负载宽度寄存器地址定义*/
#define RX_PW_P3           0x14        /*接收通道3负载宽度寄存器地址定义*/
#define RX_PW_P4           0x15        /*接收通道4负载宽度寄存器地址定义*/
#define RX_PW_P5           0x16        /*接收通道5负载宽度寄存器地址定义*/
#define FIFO_STATUS        0x17        /*FIFO 状态寄存器地址定义*/
#define DYNPD              0x1C        /*DYNPD 动态负载寄存器地址定义*/
#define FEATURE            0x1D        /*寄存器 Feature 的地址定义*/
/***************************************************************
/定义无线发送的速率
****************************************************************/
#define DATA_RATE_1M       0x00
#define DATA_RATE_2M       0x08


#define RX_P_NO            0x0e


#define MASK_IRQ_FLAGS     0x70
#define MASK_RX_DR_FLAG    0x40
#define MASK_TX_DS_FLAG    0x20
#define MASK_MAX_RT_FLAG   0x10


#define PIPE0              0x00
#define PIPE1              0x01
#define PIPE2              0x02
#define PIPE3              0x03
```

```c
#define PIPE4                   0x04
#define PIPE5                   0x05
/************************************************************************
/定义 FIFO 的状态
************************************************************************/
#define MASK_TX_EMPTY           0x10
#define MASK_RX_FULL            0x02
#define MASK_RX_EMPTY           0x01
/************************************************************************
/定义 nRF24LE1 的无线工作模式
************************************************************************/
#define Dev_Close               0
#define Dev_Normal              1
#define Dev_TX                  2
#define Dev_AutoTx              3
#define Dev_RT                  4


#ifdef __cplusplus
extern "C" {
#define extern
#endif
/************************************************************************
/声明 nRF24LE1 无线传输用到的函数原型,方便文件调用
************************************************************************/
void Rf24L01_OpenClose_Ctrl(unsigned char on_off);
void Rf24L01_Power_Ctrl(unsigned char on_off);
void Rf24L01_RxTx_Switch(unsigned char bMode);
unsigned char Rf24L01_Polling_IRQ(unsigned char * rev_buf);
unsigned char Rf24L01_Set_Init(SetupData *drc);
unsigned char Rf24L01_GetDataPipeNumber(void);
unsigned char Rf24L01_Clear_IRQ(unsigned char irq_flag);
unsigned char Rf24L01_ReadByte(unsigned char reg);
void Rf24L01_TXABYTE(unsigned char x);
void Rf24L01_Flush_TX(void);
void Rf24L01_TX(unsigned char * ptx_buf,unsigned char len);
void Rf24L01_RETX_LastFrame(void);
void Rf24L01_Set_Channel(unsigned char nChannel);
void Rf24L01_Flush_RX(void);
unsigned char Rf24L01_WriteByte(unsigned char reg, unsigned char value);
```

```c
unsigned char Rf24L01_WriteMultiByte(unsigned char reg, unsigned char * pBuf, unsigned char len);
#ifdef __cplusplus
}
#undef extern
#endif
#endif
/*********************************************************************
/                    头文件 rf24le1.h 到此结束                        /
*********************************************************************/
```

4. 定义与 MCU 操作相关的一些函数

定义与 MCU 操作相关的一些函数如下：

```c
/*********************************************************************
/文件名称：config.c
/文件说明：文件中定义了与 MCU 操作相关的一些函数,包括串口的初始化函数、I/O 口的初始化函数、
          软件延时函数、中断服务函数串口的发送/接收函数以及 I/O 口的键盘扫描函数等,用于操
          作 nRF24LE1 的单片机部分
*********************************************************************/
#include "reg24le1.h"
#include "config.h"
#include "intrins.h"
/*********************************************************************
/定义一个全局变量
*********************************************************************/
volatile unsigned char Udata;
/*********************************************************************
/函数名称：delay()
/函数功能：实现软件延时
/输入参数：x 为延时的时间数
/返回参数：无
*********************************************************************/
void delay(unsigned char x)
{
unsigned char di;
    for(;x>0;x--)
        for(di=120;di>0;di--)
```

```
        {
            ;
        }
}
/***************************************************************
/函数名称：system_init()
/函数功能：时钟初始化,射频收发器的时钟启动
/输入参数：无
/返回参数：无
***************************************************************/
void system_init(void)
{
    EA = 0;
    CLKCTRL = 0x28;
    CLKLFCTRL = 0x01;
    SPIRCON1 = 0x0B;
    RFCKEN = 1;
    WUIRQ = 0;
    EA = 1;
}
/***************************************************************
/函数名称：Uart_Init()
/函数功能：初始化 nRF24LE1 的异步串口功能,设置波特率为 2 400
/输入参数：无
/返回参数：无
***************************************************************/
void Uart_Init(void)
{
    P0DIR &= 0xF7;
    P0DIR |= 0x10;
    S0CON = 0x50;
    PCON |= 0x80;
    WDCON |= 0x80;
    S0RELL = 0x30;
    S0RELH = 0x03;
}
/***************************************************************
/函数名称：Timer_Us()
/函数功能：等待 1 μs
```

/输入参数：无
/返回参数：无
**/

```c
void Timer_Us(void)
{
    _nop_();_nop_();_nop_();_nop_();
    _nop_();_nop_();_nop_();_nop_();
    _nop_();_nop_();_nop_();_nop_();
    _nop_();_nop_();_nop_();_nop_();
    _nop_();_nop_();          _nop_();
}
```

/***
/函数名称：Timer_10Us()
/函数功能：等待指定的时间(10 μs)
/输入参数：nTime 为延时的时间数
/返回参数：无
**/

```c
void Timer_10Us(unsigned char nTime)
{
    while((nTime--)!=0) Timer_Us();
}
```

/***
/函数名称：Timer_Ms()
/函数功能：等待指定的时间(ms)
/输入参数：nTime 为延时的时间数
/返回参数：无
**/

```c
void (int nTime)
{
    int j;
    while((nTime--)!=0)
    for(j=0; j<350; j++);
}
```

/***
/函数名称：putchar()
/函数功能：通过串口发送一个字符
/输入参数：dat 为待发送的字符
/返回参数：无
**/

```c
void putchar( unsigned char dat)
{
        S0BUF = dat;
        while(!TI0);
        TI0 = 0;
}
/****************************************************************
/函数名称: puts()
/函数功能: 通过串口发送一个字符串
/输入参数: s 为指向一个字符串的指针
/返回参数: 无
****************************************************************/
void puts( char * s)
{
        while( * s != '\0')
        {
        putchar( * s);
        s ++ ;
        }
}
/****************************************************************
/函数名称: getch()
/函数功能: 读取串口接收到的一个字符
/输入参数: 无
/返回参数: 串口接收到的一个数据,否则返回 0
****************************************************************/
char getch(void)
{
char rc = 0;
if(RI0)
{
    RI0 = 0;
    rc = S0BUF;
}
 return rc;
}
/****************************************************************
/函数名称: keycheck()
/函数功能: 按键扫描函数
```

```
/输入参数:无
/返回参数:按键按下的键值
*****************************************************************/
unsigned char keycheck(void)
{
P1CON = 0xD0;           /*设置 P1.0~P1.3 以及 P1.5 为输出,其余 P1 口为输入*/
if(!P10)
{
delay(10);
if(!P10)
  { while(!P10);
    return TRUE;        /*如果按下,则返回真*/
  }
}
return FALSE;           /*返回假*/
}
/*****************************************************************
/                 C 代码文件 config.c 代码到此结束                /
*****************************************************************/
```

5. 定义用于无线传输操作的函数

定义用于无线传输操作的函数如下:

```
/*****************************************************************
/文件名称:wireless_api.c
/文件说明:在这个函数中所定义的都是用于无线传输操作的函数,包括 SPI 的读/写、多字节的读/写
         无线收发用寄存器、无线传输部分的初始化以及无线的发送和数据接收函数等。这些都是
         无线传输操作的核心函数
/*nRF24LE1 无线部分的初始化参数配置函数定义*/
/*****************************************************************
/定义一些简单的函数
*****************************************************************/
#include "reg24le1.h"
#include "wireless_api.h"
#include "rf24le1.h"
/*****************************************************************/
#define CSN_LOW()        RFCSN = 0;
#define CSN_HIGH()       RFCSN = 1;
```

```c
#define CE_LOW()        RFCE = 0;Timer_10Us(1);
#define CE_HIGH()       RFCE = 1;
#define CE_PULSE()      CE_HIGH();Timer_10Us(2);CE_LOW();
/****************************************************************
/用到的全局变量的定义
*****************************************************************/
idata unsigned char radio_status;                    /*无线状态*/
static unsigned char gnDataPipeNumber;               /*无线通信的数据通道号*/
code const address[] = {0x12,37,0x55,0x79,0x97};     /*无线收发地址设置*/
/****************************************************************
/函数名称：Rf24L01_Reg_RW()
/函数功能：nRF24LE1的无线部分SPI接口读/写操作
/输入参数：byte为SPI接口要传输的一个数据
/返回参数：之前的状态值
*****************************************************************/
static unsigned char Rf24L01_Reg_RW(unsigned char byte)
{
    SPIRDAT = byte;                  /*写入要传送的一字节*/
    RFSPIF = 0;                      /*清除标志*/
    while(!RFSPIF);                  /*等待完成*/
    return SPIRDAT;                  /*返回状态*/
}
/****************************************************************
/函数名称：Rf24L01_WriteByte()
/函数功能：给指定无线配置寄存器中写入指定的值
/输入参数：reg为寄存地址,value为要写入的值
/返回参数：reg中原来的值
*****************************************************************/
unsigned char Rf24L01_WriteByte(unsigned char reg, unsigned char value)
{
    unsigned char status;
    CSN_LOW();
    status = Rf24L01_Reg_RW(reg);    /*选定要写入的寄存器*/
    Rf24L01_Reg_RW(value);           /*写入值*/
    CSN_HIGH();
    return(status);                  /*返回寄存器状态*/
}
/****************************************************************
```

```
/*函数名称：Rf24L01_WriteMultiByte()
/*函数功能：指定的寄存器,并一次性连续写入多个值
/*输入参数：reg 为寄存器地址,pBuf 为一个数据指针,len 为写入数据的个数
/*返回参数：起始值
******************************************************************/
unsigned char Rf24L01_WriteMultiByte(unsigned char reg, unsigned char * pBuf, unsigned char len)
{
    unsigned char status,i;
    CSN_LOW();
    status = Rf24L01_Reg_RW(reg);              /*选定要写的寄存器*/
    for(i = 0; i<len; i++)                      /*将要写的值写入缓冲区*/
    {
        Rf24L01_Reg_RW( * pBuf ++);
    }
    CSN_HIGH();
    return(status);                             /*返回状态位*/
}
/******************************************************************
/*函数名称：Rf24L01_Clear_IRQ()
/*函数功能：清除无线中断请求标志
/*输入参数：flag 为用于清除中断的标志
/*返回参数：传输状态值
******************************************************************/
unsigned char Rf24L01_Clear_IRQ(unsigned char irq_flag)
{
    return Rf24L01_WriteByte(WRITE_REG + STATUS, irq_flag);
}
/******************************************************************
/*函数名称：Rf24L01_Flush_TX()
/*函数功能：刷新无线发送缓冲区
/*输入参数：无
/*返回参数：无
******************************************************************/
void Rf24L01_Flush_TX(void)
{
    Rf24L01_WriteByte(FLUSH_TX,0);
}
/******************************************************************
```

```
/函数名称: Rf24L01_Flush_RX()
/函数功能: 刷新无线接收缓冲区
/输入参数: 无
/返回参数: 无
*********************************************************/
void Rf24L01_Flush_RX(void)
{
    Rf24L01_WriteByte(FLUSH_RX,0);
}
/*********************************************************
/函数名称: Rf24L01_ReadByte()
/函数功能: 读取一个寄存器的值
/输入参数: reg 为待读的寄存器地址
/返回参数: 读取的寄存器的值
*********************************************************/
unsigned char Rf24L01_ReadByte(unsigned char reg)
{
    unsigned char reg_val;
    CSN_LOW();
    Rf24L01_Reg_RW(reg);                    /*选定要读的寄存器*/
    reg_val = Rf24L01_Reg_RW(0);            /*读取寄存器值*/
    CSN_HIGH();
    return(reg_val);                        /*返回寄存器值*/
}
/*********************************************************
/函数名称: Rf24L01_RxTx_Switch()
/函数功能: 实现无线的发送和接收模式的切换
/输入参数: bMode 为待设定的无线的工作模式
/返回参数: 无
*********************************************************/
void Rf24L01_RxTx_Switch(unsigned char bMode)
{
    unsigned char bConfig;
    CE_LOW();
    Rf24L01_Flush_RX();                     /*清空收发缓冲区*/
    Rf24L01_Flush_TX();
    Rf24L01_Clear_IRQ(MASK_IRQ_FLAGS);
    bConfig = Rf24L01_ReadByte(CONFIG);     /*读取当前无线收发用配置寄存器的值*/
```

```c
    if(bMode == PRX)
    {
        if((bConfig&0x01))
        {
            CE_HIGH();
            return;
        }
        bConfig &= 0xfe;
        bConfig |= 0x01;
        Rf24L01_WriteByte(WRITE_REG + CONFIG, bConfig);
        CE_HIGH();
    }
    else if(bMode == PTX)
    {
        if(!(bConfig&0x01))
        {
            return;
        }
        bConfig &= 0xfe;
        Rf24L01_WriteByte(WRITE_REG + CONFIG, bConfig);
    }
}
/****************************************************************
/函数名称：Rf24L01_ReadMultiByte()
/函数功能：从一个寄存器地址连续读出 len 个数据，存入 pBuf 指向的存储单元
/输入参数：reg 为寄存器地址，pBuf 为指向存储单元的指针，len 为数据个数
/返回参数：状态量
****************************************************************/
static unsigned char Rf24L01_ReadMultiByte(unsigned char reg, unsigned char * pBuf, unsigned char len)
{
    unsigned char status,i;
    CSN_LOW();
    status = Rf24L01_Reg_RW(reg);            /*选定操作的寄存器*/
    for(i = 0;i<len;i++)
    {
        pBuf[i] = Rf24L01_Reg_RW(0);         /*读取数据*/
    }
```

```c
    pBuf[i] = '\0';
    CSN_HIGH();
    return(status);                              /*返回状态字*/
}
/****************************************************************
/函数名称: Rf24L01_RX()
/函数功能: 用于接收有效的无线数据
/输入参数: prx_bug 为指向数据存储单元的指针
/返回参数: 无
****************************************************************/
static void Rf24L01_RX(unsigned char * prx_buf)
{
    unsigned char gnRF_RecvLen;
    gnRF_RecvLen = Rf24L01_ReadByte(READ_PAYLAODLEN);    /*读取接收数据长度*/
    if (prx_buf != NULL)
Rf24L01_ReadMultiByte(RD_RX_PLOAD,prx_buf,gnRF_RecvLen); /*读取全部数据*/
}
/****************************************************************
/函数名称: Rf24L01_TX()
/函数功能: 利用无线发送一个字符串
/输入参数: ptx_buf 为指向待发送的数据单元的指针,nLen 为待发送的数据个数
/返回参数: 无
****************************************************************/
void Rf24L01_TX(unsigned char * ptx_buf,unsigned char nLen)
{
    Rf24L01_RxTx_Switch(PTX);                    /*转换到发送模式*/
    Rf24L01_Flush_TX();                          /*清空发送缓冲区*/
    Rf24L01_WriteMultiByte(WR_TX_PLOAD, ptx_buf, nLen);  /*连续写入要发送的数据*/
    CE_PULSE();                                  /*20 μs 的高脉冲,激发数据发送*/
}
/****************************************************************
/函数名称: Rf24L01_TXABYTE()
/函数功能: nRF24LE1 无线发送一字节的数据
/输入参数: x 为待发送的一字节的数据
/返回参数: 无
****************************************************************/
void Rf24L01_TXABYTE(unsigned char x)
{
```

```c
        Rf24L01_RxTx_Switch(PTX);                      /*转换到发送模式*/
        Rf24L01_Flush_TX();                            /*清除发送缓冲区*/
        Rf24L01_WriteByte(WR_TX_PLOAD, x);             /*写入要发送的数据*/
        CE_PULSE();                                    /*20 μs 的高脉冲*/
}
/*********************************************************************
/中断服务处理函数,用于处理无线传输中的一些情况
**********************************************************************/
static void rf_rdy_ov_interrupt(void) interrupt INTERRUPT_RFRDY {}
static void rfirq_ov_interrupt(void) interrupt INTERRUPT_RFIRQ
{
        radio_status = 0xFF;
}
/*********************************************************************
/函数名称：Rf24L01_Polling_IRQ()
/函数功能：通过扫描中断的标志的方式接收无线发来的数据
/输入参数：rev_buf 为指向数据存储单元的指针
/返回参数：当前数据状态
**********************************************************************/
unsigned char Rf24L01_Polling_IRQ(unsigned char * rev_buf)
{
        unsigned char irq_status;
        if(radio_status == 0x00)
        {
                if(Rf24L01_ReadByte(FIFO_STATUS) & MASK_RX_EMPTY) return IDLE;
                Rf24L01_RX(rev_buf);                   /*将收到的有效数据存到revbuf指针当中*/
                irq_status = Rf24L01_ReadByte(STATUS); /*读取 SPI 状态寄存器*/
                gnDataPipeNumber = (irq_status >> 1)&0x7;/*取当前收到数据的通道号*/
                return (unsigned char)RX_DR;           /*成功地接收到数据*/
        }
        radio_status = 0x00;
        irq_status = Rf24L01_ReadByte(STATUS);         /*读无线收发用状态寄存器*/
        switch(irq_status&MASK_IRQ_FLAGS)              /*取 4、5、6 位*/
        {
                case MASK_RX_DR_FLAG:                  /*接收到数据*/
                {
                        if(Rf24L01_ReadByte(FIFO_STATUS) & MASK_RX_EMPTY)/*接收缓冲区空*/
                        {
```

```c
            irq_status = IDLE;
        }
        else
        {
            Rf24L01_RX(rev_buf);
            gnDataPipeNumber = (irq_status >> 1)&0x7;
            irq_status = (unsigned char)RX_DR;

        }
        break;
    }
    case MASK_TX_DS_FLAG:                        /*带自动应答的数据发送成功*/
    {
        irq_status = (unsigned char)TX_DS;
        break;
    }
    case MASK_MAX_RT_FLAG:                       /*最大重发触发了中断*/
    {
        irq_status = (unsigned char)MAX_RT;
        break;
    }
    case IDLE:                                   /*没有数据*/
    {
        irq_status = (unsigned char)IDLE;
        break;
    }
    }
    if(Rf24L01_ReadByte(FIFO_STATUS) & MASK_RX_EMPTY)  /*接收空,清除标志*/
        Rf24L01_Clear_IRQ(MASK_IRQ_FLAGS);
    return irq_status;
}
/*************************************************************
/函数名称：Rf24L01_Set_Init()
/函数功能：初始化 nRF24LE1 无线收发功能
/输入参数：drc 为指向 SetupData 类型的一个无线配置结构体的指针
/返回参数：一个配置是否完成的一个标志量
**************************************************************/
```

第 4 章 nRF24LE1 的射频收发器与应用

```c
unsigned char Rf24L01_Set_Init(SetupData * drc)
{
    unsigned char btemp;
    RFCKEN = 1;                                    /* 使能无线传输时钟 */
    RF = 0;                                        /* RF 中断清零 */
    CE_LOW();                                      /* 选中 RF */
    Rf24L01_Flush_RX();                            /* 清空接收发送缓冲 */
    Rf24L01_Flush_TX();
    Rf24L01_Clear_IRQ(MASK_IRQ_FLAGS);             /* 清除中断请求状态寄存器 */
    /*************************************************************
    /设置无线的配置寄存器,这个很重要
    *************************************************************/
    btemp = 0x0e;
    if (drc->nOn == Dev_Normal) btemp += 0x01;
    Rf24L01_WriteByte(WRITE_REG + CONFIG, btemp);
    /*************************************************************
    /设置通道自动应答
    *************************************************************/
    Rf24L01_WriteByte(WRITE_REG + EN_AA, drc->nAutoAck);
    /*************************************************************
    /使能接收通道地址配置
    *************************************************************/
    Rf24L01_WriteByte(WRITE_REG + EN_RXADDR, drc->nRecvAddr);
    /*************************************************************
    /设置无线模块地址宽度
    *************************************************************/
    Rf24L01_WriteByte(WRITE_REG + SETUP_AW, 0x03);
    /*************************************************************
    /设置重传延时和最多重传的次数
    *************************************************************/
    btemp = (drc->nART_Factor&0x0f) << 4;
    btemp += (drc->nRetran > 15 ? 15 : drc->nRetran);
    btemp = 0x17;//self
    Rf24L01_WriteByte(WRITE_REG + SETUP_RETR, btemp);
    /*************************************************************
    /设置无线收发的工作频点
    *************************************************************/
    if (drc->nChannel>122)
    {
        drc->nChannel = 0;
```

```c
    }
    Rf24L01_WriteByte(WRITE_REG + RF_CH, drc->nChannel);
/*****************************************************************
/设置无线收发的传输速率和传输功率
******************************************************************/
    btemp = ((drc->nPower&0x0f)>3 ? 3:(drc->nPower&0x0f)) << 1;
    btemp += 0x01;
    if((drc->nPower&0xf0)>0) btemp += 0x08;
    btemp = 0x0e;
    Rf24L01_WriteByte(WRITE_REG + RF_SETUP, btemp);
/*****************************************************************
/设置通道 0 地址和数据宽度
******************************************************************/
    Rf24L01_WriteMultiByte(WRITE_REG + RX_ADDR_P0,drc->aAddr1,5);
    if (drc->nLen>32)
    {
        drc->nLen = 32;
    }
    else if (drc->nLen == 0)
    {
        drc->nLen = 1;
    }
    Rf24L01_WriteByte(WRITE_REG + RX_PW_P0, drc->nLen);
    btemp = drc->nRecvAddr;
    btemp = btemp >> 1;
    if (btemp&1)
    {
/*****************************************************************
/设置接收通道 1 的地址
******************************************************************/
        Rf24L01_WriteMultiByte(WRITE_REG + RX_ADDR_P1,drc->aAddr2,5);
/*****************************************************************
/设置接收通道 1 的数据宽度
******************************************************************/
        Rf24L01_WriteByte(WRITE_REG + RX_PW_P1, drc->nLen);
    }
/*****************************************************************
/设置无线收发的地址
******************************************************************/
    Rf24L01_WriteMultiByte(WRITE_REG + TX_ADDR,drc->aAddr1,5);
```

第4章 nRF24LE1 的射频收发器与应用

```
/*******************************************************************
/设置无线收发用的特征寄存器
*******************************************************************/
    Rf24L01_WriteByte(WRITE_REG + FEATURE, 0x05);
/*******************************************************************
/设置无线收发的动态数据长度
*******************************************************************/
    Rf24L01_WriteByte(WRITE_REG + DYNPD, drc ->nRecvAddr);
    RF = 1;
    if(drc ->nOn == Dev_Normal)
    {
        CE_HIGH();
    }
    return 0;
}
/*******************************************************************
/函数名称：wireless_init()
/函数功能：nRF24LE1 的无线部分的总初始化程序
/输入参数：无
/返回参数：无
*******************************************************************/
void wireless_init(void)
{
    SetupData mSetup;
    memset(&mSetup,0,sizeof(SetupData));      /*将结构体 SetupData 初始化为 0*/
    mSetup.nChannel = 20;
    mSetup.nPower = 3;                         /*功率设置成 0 dB*/
    mSetup.nRecvAddr = 0x01;
    mSetup.nAutoAck = 0x01;
    mSetup.nLen = 32;
    mSetup.nRetran = 8;
    mSetup.nOn = 1;
    memcpy(mSetup.aAddr1,address,5);           /*address 地址复制到前面一个空间*/
    mSetup.nOn = 1;

    Rf24L01_Set_Init((SetupData *)&mSetup);    /*调用结构体初始化无线部分*/
}
/*******************************************************************
/                   C 代码文件 wireless_api.c 到此结束              /
*******************************************************************/
```

6. 无线传输示例程序的主函数

无线传输示例程序的主函数如下：

```c
/**************************************************************
/C语言文件三：wireless.c
/文件说明：本文件是该示例程序的主函数文件,所有的无线传输用的函数都在这里调用,主函数
          是程序的入口,通过在主函数中调用定义的无线功能函数可以实现无线通信
**************************************************************/
#include "reg24le1.h"
#include "config.h"
#include "wireless_api.h"
#include "rf24le1.h"
/**************************************************************
/宏定义一个量用于后面的条件编译
/屏蔽此宏定义最终编译出来的HEX文件作为接收端固件程序
/执行此宏定义最终编译出来的HEX文件作为发送端固件程序
**************************************************************/
#define SEND_24LE1
/**************************************************************
/定义主函数中用到的一些变量
**************************************************************/
unsigned char flag = 0;
unsigned char buffer[32];
unsigned char text[1] = "\0";
/**************************************************************
/主函数
**************************************************************/
void main(void)
{
    system_init();
    Uart_Init();
    wireless_init();
    #ifdef SEND_24LE1                  /*如果定义SEND_24LE1,则设置成无线发送模式*/
    Rf24L01_RxTx_Switch(PTX);
    #else
    Rf24L01_RxTx_Switch(PRX);          /*否则设置成无线接收模式*/
    #endif
    puts("无线测试程序,波特率为2400!\n");
```

```
while(1)
{
    #ifdef SEND_24LE1              /*如果是定义发送,那么条件编译发送端*/
    if(RI0)
    {
    RI0 = 0;
    text[0] = S0BUF;
    Rf24L01_TX((unsigned char *)text,1);   /*发送接收的字符*/
    S0BUF = text[0];
    while(!TI0);                   /*回显用于监视*/
    TI0 = 0;
    }
    #else                          /*条件编译接收端*/
    if (Rf24L01_Polling_IRQ((unsigned char *)buffer) == RX_DR)  /*接收字符*/
    {
        puts(buffer);              /*串口显示字符*/
    }
    #endif                         /*结束条件编译*/
}
```

/***
/   C代码文件 wireless.c 到此结束   /
***/

7. 无线传输示例程序运行结果

整个程序采用条件编译,通过检查 SEND_24LE1 是否被宏定义来决定程序是编译成发送端还是接收端的固件程序。

在本示例中,将工程两次编译后生成的发送端固件程序和接收端固件程序分别下载到无线串口的发送端和接收端,再将发送端串口和接收端串口分别连接在两台计算机的串口上,上电测试无线串口的数据传输功能。数据通过与发送端串口相连的计算机上的串口调试助手发出,在有效接收距离内可在与接收端串口相连的计算机超级终端软件上接收到,接收到的数据与发送端发送的数据一致。

测试结果如图 4.2.4 和图 4.2.5 所示,其中图 4.2.4 是发送端串口连接着的计算机上运行的串口调试助手的截图,图 4.2.5 是接收端串口连接着的计算机上的超级终端软件的截图。从两个截图上可以看到,发送端发出的数据已经被接收端正确地接收。

第 4 章 nRF24LE1 的射频收发器与应用

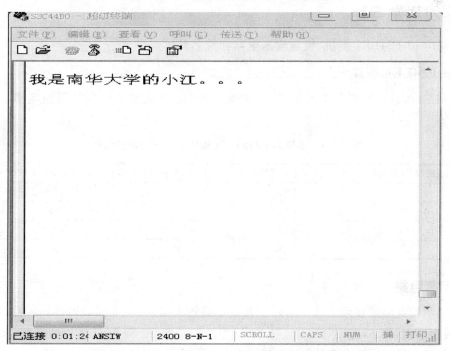

图 4.2.4　发送端计算机显示器截图

图 4.2.5　接收端计算机显示器截图

第4章 nRF24LE1 的射频收发器与应用

4.3 射频收发器应用示例 2

4.3.1 系统结构

系统由发送端和接收端两部分组成。发送端由 nRF24LE1 电路模块、按键输入电路和液晶显示器组成;接收端由 nRF24LE1 模块、MP3 语音模块、音频功率放大器电路组成。系统结构方框图如图 4.3.1 所示。系统利用发送端按键控制接收端的 MP3 音乐播放器工作。

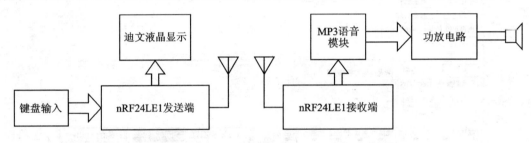

图 4.3.1 系统结构方框图

4.3.2 发送端电路

发送端由 nRF24LE1 模块、键盘输入电路和液晶显示器组成。

1. nRF24LE1 模块

nRF24LE1 模块电路请参考 1.2 节。发送端 nRF24LE1 模块 I/O 与外设的连接如表 4.3.1 所列。

表 4.3.1 发送端 nRF24LE1 模块 I/O 与外设的连接

nRF24LE1 模块的 I/O	发送端外设网络标号	nRF24LE1 模块的 I/O	发送端外设网络标号
P1.0	I/O1	P1.4	I/O5
P1.1	I/O2	P0.7	I/O6
P1.2	I/O3	P0.4	TXD
P1.3	I/O4	P0.3	RXD

2. 按键电路

发送端的按键电路如图 4.3.2 所示。按键电路用来实现 MP3 播放器的相关控制。例如,按 S1 播放上一曲,按 S2 播放下一曲,按 S3 增加播放音量,按 S4 降低播放的音量,按 S5 实现暂停和恢复播放,LED(D0)用来指示按键状态。

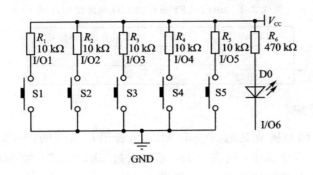

图 4.3.2　按键电路

3. 液晶显示器

液晶显示器选用北京迪文科技有限公司的 T600 串口液晶显示器。T600 液晶显示器自带中英文字库和大容量的存储器，支持迪文 HMI 指令集，通过串口发送指令，即可在显示屏上的指定位置显示图片、文字、曲线、进度条等内容。迪文 HMI 指令集请登录北京迪文科技有限公司的网站查询。

nRF24LE1 模块与液晶显示器的连接电路如图 4.3.3 所示。

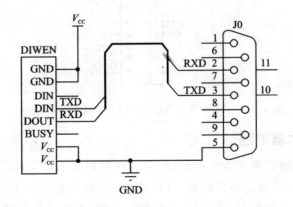

图 4.3.3　nRF24LE1 模块与液晶显示器的连接

4.3.3　接收端电路

接收端由 nRF24LE1 模块、MP3 语音模块和音频功率放大器电路组成。

1. nRF24LE1 模块

nRF24LE1 模块电路请参考 1.2 节。接收端 nRF24LE1 模块 I/O 与外设的连接如表 4.3.2 所列。

表 4.3.2　nRF24LE1 接收端 I/O 与外设的连接

接收端 nRF24LE1 的 I/O	接收端外设网络标号
P0.4	TXD
P0.3	RXD

2. MP3 语音模块

接收端的语音解码和播放是通过一个 MP3 语音模块实现的，MP3 语音模块采用 SK-SDMP3_V1.7 通用工业级 MP3 语音模块，可以通过 nRF24LE1 的 MCU 串口命令控制。MCU 可以通过串口命令控制插接在 MP3 语音模块上的 SD 卡里的 MP3 文件。MP3 语音模块在控制命令的操作下，可以按照需要读取 SD 卡内的音乐文件进行播放，实现暂停、恢复、重放和随机播放，也可以实现音量的增加、降低、静音、下一曲和上一曲等功能。

MP3 语音模块与接收端 nRF24LE1 模块的连接如图 4.3.4 所示。

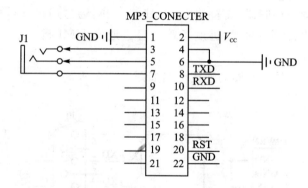

图 4.3.4　MP3 语音模块与 nRF24LE1 模块的连接

3. 音频功率放大器电路

音频功率放大器电路如图 4.3.5 所示，功放芯片采用 TDA2030。TDA2030 是飞利浦公司生产的音频功放模块，采用超小型封装（TO-220），输出功率 $P_O = 18\ W(R_L = 4\ \Omega)$，电源电压为 $\pm 6 \sim \pm 22\ V$，具有短路保护、热保护、地线偶然开路、电源极性反接及负载泄放电压反冲等保护电路。

4. ±5 V 和 ±12 V 电源电路

电源电路如图 4.3.6 所示，图中 AC1 和 AC2 连接 18 V 电源变压器的两个输出端，GND 接电源变压器的公共地端。电路可以提供 ±5 V 和 ±12 V 供电。nRF24LE1 模块的供电电压是 +3.3 V，可以将 +5 V 电源电压通过稳压芯片 ASM1117-3.3 稳压后提供。

第 4 章 nRF24LE1 的射频收发器与应用

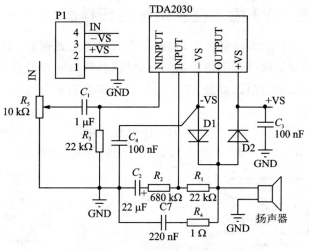

图 4.3.5 音频功率放大器电路

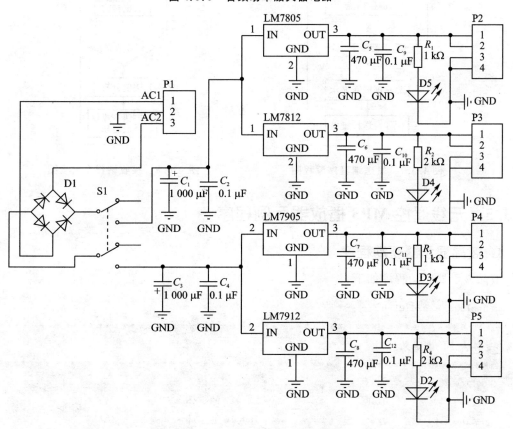

图 4.3.6 ±5 V 和 ±12 V 电源电路

4.3.4 无线遥控 MP3 播放器示例程序流程图

本示例以 nRF24LE1 作为控制核心，实现一个简单的无线遥控 MP3 播放器功能。通过操作发送端的按键，实现对接收端的播放器的有效控制，例如实现暂停、恢复、重放、随机播放功能，以及实现音量的增加、降低、静音、下一曲和上一曲等功能。同时，在发送端的液晶显示器上显示对应的操作信息。

该系统分为发送端和接收端。发送端的程序流程图如图 4.3.7 所示，接收端程序的流程图如图 4.3.8 所示。由于示例程序代码较长，为了使程序的层次清晰，分成了多个文件。

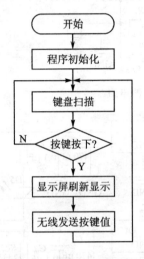

图 4.3.7 发送端程序流程图

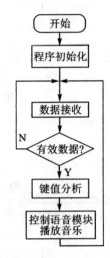

图 4.3.8 接收端程序流程图

4.3.5 无线遥控 MP3 播放器示例程序

1. 液晶显示器驱动程序

液晶显示器驱动程序如下：

```
#include "reg24le1.h"
/******************************************************
/字体大小的宏定义
******************************************************/
#define F8      0x53      /*8×8 的字体*/
#define F12     0x6E      /*12×12 的字体*/
#define F16     0x54      /*16×16 的字体*/
#define F24     0x6F      /*24×24 的字体*/
```

```
#define F32        0x55          /* 32×32 的字体 */
#define head       0xAA
```
/**
/重定义数据类型
**/
```
typedef unsigned char uchar;
typedef unsigned int uint;
```
/**
/函数名称：uartsendB()
/函数功能：利用串口发送一字节
/输入参数：dat 为待发送的字符
/返回参数：无
**/
```
void uartsendB(unsigned char dat)
{
    S0BUF = dat;
    while(!TI0);
    TI0 = 0;
}
```
/**
/函数名称：uartsendS()
/函数功能：利用串口发送字符串
/输入参数：p 为指向待发送的字符的指针，s 为发送的数据字节数
/返回参数：无
**/
```
void uartsendS(uchar * p,uchar s)
{
    uchar m;
    for(m = 0;m<s;m++)
    {
        uartsendB( * p);
        p++;
    }
}
```
/**
/函数名称：DwinW()
/函数功能：往液晶显示模块写入一个字长的数据
/输入参数：x 为待传输给液晶模块的数据
/返回参数：无

```c
/*****************************************************************/
void DwinW(uint x)
{
    uartsendB(x >> 8);
    uartsendB(x);
}
/*****************************************************************
/函数名称：uartend()
/函数功能：发送液晶模块控制指令结束标志
/输入参数：无
/返回参数：无
*****************************************************************/
void uartend()
{
    uartsendB(0xcc);
    uartsendB(0x33);
    uartsendB(0xc3);
    uartsendB(0x3c);
}
/*****************************************************************
/函数名称：uartpos()
/函数功能：串口传输两个字给液晶
/输入参数：x、y为两个字
/返回参数：无
*****************************************************************/
void uartpos(uint x,uint y)
{
    uartsendB(x >> 8);
    uartsendB(x);
    uartsendB(y >> 8);
    uartsendB(y);
}
/*****************************************************************
/函数名称：uartp()
/函数功能：根据参数锁定坐标点到合适的位置显示指定大小的字符
/输入参数：x和y分别为屏幕横坐标和纵坐标,font为显示的阵列类型
/返回参数：无
*****************************************************************/
void uartp(uchar x,uchar y,char font)
```

```
{
    x--;
    y--;
    if(font == F8)
        uartpos(x << 3,y << 3);
    else if(font == F12)
        uartpos((x << 3)+(x << 2),(y << 3)+(y << 2));
    else if(font == F16)
        uartpos((x << 4),(y << 4));
    else if(font == F24)
        uartpos((x << 4)+(x << 3),(y << 4)+(y << 3));
    else if(font == F32)
        uartpos((x << 5),(y << 5));
}
/****************************************************************
/函数名称：uartp1()
/函数功能：调整字符显示的坐标到合适的位置
/输入参数：x、y 和 font
/返回参数：无
****************************************************************/
void uartp1(uchar x,uchar y,char font)
{
    x--;
    y--;
    if(font == F8)
        uartpos(x << 2,y << 3);
    else if(font == F12)
        uartpos((x << 2)+(x << 1),(y << 3)+(y << 2));
    else if(font == F16)
        uartpos((x << 3),(y << 4));
    else if(font == F24)
        uartpos((x << 3)+(x << 2),(y << 4)+(y << 3));
    else if(font == F32)
        uartpos((x << 4),(y << 5));
}
/****************************************************************
/函数名称：Dwinchar()
/函数功能：在液晶指定坐标显示一个指定大小的字符
/输入参数：x、y 为显示的坐标点，font 为文字大小，byte 为显示内容
```

/返回参数:无
**/
```c
void Dwinchar(uchar x,uchar y, char font, char byte)
{
    uartsendB(head);
    uartsendB(font);
    uartsendB(x/256);
    uartsendB(x % 256);
    uartsendB(y/256);
    uartsendB(y % 256);
    uartsendB(byte);
    uartend();
}
```
/**
/函数名称:prints()
/函数功能:液晶上显示一串字符
/输入参数:x、y为显示字符的起始坐标,font为尺寸,s为字符指针
/返回参数:无
**/
```c
void prints(int x,int y,uchar font , uchar * s)
{
    uartsendB(head);            /* 帧头 0xAA */
    uartsendB(font);            /* 汉字大小 */
    uartsendB(x/256);
    uartsendB(x % 256);
    uartsendB(y/256);
    uartsendB(y % 256);
    while( * s != '\0')         /* 发送汉字字串,直到发送完全部字符 */
    {
        uartsendB( * s);
        s++;
    }
    uartend();                  /* 发送帧结束符 */
}
```
/**
/函数名称:DwinON()
/函数功能:打开或者关闭液晶显示模块背景灯
/输入参数:x为液晶模块背景灯开关标志
/返回参数:无

```
*****************************************************/
void DwinON(char x)
{
    uartsendB(head);
    if(x)
        uartsendB(0x5e);
    else
        uartsendB(0x5f);
    uartend();
}
/****************************************************
/函数名称：DwinColor()
/函数功能：设置液晶显示模块调色板
/输入参数：Fcolor 为前景色，Bcolor 为背景色
/返回参数：无
*****************************************************/
void DwinColor(unsigned int Fcolor,unsigned int Bcolor)
{
    uartsendB(head);
    uartsendB(0x40);
    uartsendB(Fcolor >> 8);
    uartsendB(Fcolor);
    uartsendB(Bcolor >> 8);
    uartsendB(Bcolor);
    uartend();
}
/****************************************************
/函数名称：DwinFcolor()
/函数功能：取指定的颜色到调色板
/输入参数：x、y 为取点的坐标，font 为取点的大小
/返回参数：无
*****************************************************/
void DwinFcolor(uchar x,uchar y,uchar font)
{
    uartsendB(head);
    uartsendB(0x42);
    uartp(x,y,font);
    uartend();
}
```

```c
/***************************************************************
/函数名称：DwinColor1()
/函数功能：设置液晶显示器前景色
/输入参数：Fcolor 为前景色颜色参数
/返回参数：无
***************************************************************/
void DwinColor1(uint Fcolor)
{
    uartsendB(head);
    uartsendB(0x42);
    uartsendB(Fcolor >> 8);
    uartsendB(Fcolor);
    uartend();
}

/***************************************************************
/函数名称：DwinClear()
/函数功能：显示屏清屏，显示设置的背景色
/输入参数：无
/返回参数：无
***************************************************************/
void DwinClear(void)
{
    uartsendB(head);
    uartsendB(0x52);
    uartend();
}

/***************************************************************
/函数名称：DwinPicture()
/函数功能：切换显示屏显示的图片
/输入参数：num 为存储在液晶模块中图片的编号
/返回参数：无
***************************************************************/
void DwinPicture(char num)
{
    uartsendB(head);
    uartsendB(0x70);
    uartsendB(num);
    uartend();
}
```

```
/****************************************************************
/函数名称：DwinCursor()
/函数功能：在指定的坐标点,显示指定大小尺寸的光标
/输入参数：font 为点阵类型,cursorEn 为显示光标的开关,x 为显示的横坐标,y 为显示的纵坐标,cur-
          sorWidth 为显示光标的像素宽度,cursorHeight 为光标高度
/返回参数：无
****************************************************************/
void DwinCursor(uchar font,uchar cursorEn,uint x ,uint y,uchar cursorWidth,uchar cursorHeight)
{
    uartsendB(head);
    uartsendB(0x44);
    uartsendB(cursorEn);
    uartp1(x,y,font);
    uartsendB(cursorWidth);
    uartsendB(cursorHeight);
    uartend();
}
/****************************************************************
/函数名称：Dwinrec()
/函数功能：在液晶屏上绘制一个矩形
/输入参数：x1、y1、x2、y2 分别为矩形的左上和右下坐标
/返回参数：无
****************************************************************/
void Dwinrec(uint x1,uint y1,uint x2,uint y2)
{
    uartsendB(head);
    uartsendB(0x59);
    uartpos(x1,y1);
    uartpos(x2,y2);
    uartend();
}

/****************************************************************
/函数名称：Dwinfillw()
/函数功能：用颜色填充指定的矩形区域
/输入参数：x、y、x1、y1 分别为填充矩形的左上和右下坐标
/返回参数：无
****************************************************************/
void Dwinfillw(uint x,uint y, uint x1, uint y1)
```

```
{
    uartsendB(head);
    uartsendB(0x5B);
    uartpos(x,y);
    uartpos(x1,y1);
    uartend();
}
/***************************************************************
/函数名称：DwinJingdu()
/函数功能：开机界面的进度条绘制函数，同时显示一些文字信息
/输入参数：x、y为开机进度条的左上角坐标
/返回参数：无
***************************************************************/
void DwinJingdu(uint x,uint y,uint step)
{
    prints(70,40,F24,"系统启动。。。");
    if(step == 199)
    {step = 200;
    Dwinchar(x + 240,y - 20,F16,(step/2)/100 + '0');
    }
    Dwinchar(x + 252,y - 20,F16,((step/2) % 100)/10 + '0');
    Dwinchar(x + 264,y - 20,F16,(step/2) % 10 + '0');
    Dwinchar(x + 276,y - 20,F16,'%');

    DwinColor1(0x001f);
    Dwinrec(x - 2,y - 2,x + 202,y + 14);
    Dwinrec(x,y,x + 200,y + 12);
    DwinColor(0x07e0,0x00);
    Dwinfillw(x + 1,y + 1,x + 1 + step,y + 11);
    Dwinclrw(x + 2 + step,y + 1,x + 200,y + 11);
    prints(200,200,F16,"小江制作");
}
/***************************************************************
/函数名称：baudinint()
/函数功能：初始化 nRF24LE1 的串口
/输入参数：baud 为准备设置的串口波特率
/返回参数：无
***************************************************************/
void baudinint(unsigned int baud)
```

```c
{
    CLKCTRL = 0x28;
    CLKLFCTRL = 0x01;
    P0DIR &= 0xF7;              /*配置P0.3(TXD)为输出*/
    P0DIR |= 0x10;              /*配置P0.4(RXD)为输入*/
    P0 |= 0x18;
    S0CON = 0x50;
    PCON |= 0x80;               /*设置波特率倍增*/
    WDCON |= 0x80;              /*选定内部波特率发生器*/
    if(baud == 38400)
    {
        S0RELL = 0xF3;          /*设置波特率为38 400*/
        S0RELH = 0x03;
    }
    else if(baud == 9600)
    {
        S0RELL = 0xCC;          /*设置波特率为9 600*/
        S0RELH = 0x03;
    }
}
/****************************************************************
/函数功能: dispword()
/函数功能: 显示指定的一个字符串
/输入参数: 无
/返回参数: 无
****************************************************************/
void dispword(void)
{
    static uchar n = 1;
    prints(100,100,F32,"南华大学!");
    n++;
    n = n%5;
    if(!n)
    n = 1;
    DwinON(0);                  /*背景灯关闭*/
}
/****************************************************************
/函数名称: showstop()
/函数功能: 屏幕显示处于暂停提示字符
/输入参数: 无
```

/返回参数：无
***/

```c
void showstop(void)
{
prints(100,180,F32,"Stopping...");
}
```
/**
/函数名称：showplay()
/函数功能：屏幕显示正在播放提示字符
/输入参数：无
/返回参数：无
***/

```c
void showplay(void)
{
prints(100,180,F32,"Playing...");
}
```
/**
/函数名称：showvol()
/函数功能：屏幕显示当前播放器的音量
/输入参数：x为当前播放器的放音音量
/返回参数：无
***/

```c
void showvol(char x)
{
prints(10,10,F16,"VOL:");
Dwinchar(50,10,F16,x+'0');
}
```
/**
/函数名称：shownext()
/函数功能：屏幕显示下一曲提示字符串
/输入参数：无
/返回参数：无
***/

```c
void shownext(void)
{
prints(100,180,F32,"Nextone...");
}
```
/**
/函数名称：showback()
/函数功能：屏幕显示上一曲提示字符串

/输入参数：无
/返回参数：无
**/
void showback(void)
{
prints(100,180,F32,"Lastone...");
}
/**
/函数名称：showtitle()
/函数功能：播放器主界面的文字显示，包括曲目和播放时间
/输入参数：num 为曲目，m 为播放时间的分钟，s 为播放时间的秒
/返回参数：无
**/
void showtitle(unsigned char num,unsigned char m,unsigned char s)
{
prints(25,40,F32,"Easy Music Player");
prints(60,100,F16,"Name:");
Dwinchar(110,100,F16,num/100 + '0');
Dwinchar(120,100,F16,(num%100)/10 + '0');
Dwinchar(130,100,F16,num%10 + '0');
prints(140,100,F16,".mp3");
prints(60,130,F16,"Time:");
Dwinchar(110,130,F16,m/10 + '0');
Dwinchar(120,130,F16,m%10 + '0');
Dwinchar(130,130,F16,':');
Dwinchar(140,130,F16,s/10 + '0');
Dwinchar(150,130,F16,s%10 + '0');
}
/**
/   迪文液晶显示器的驱动程序到此结束   /
**/

2. MP3 语音模块控制程序

MP3 语音模块的控制程序如下：

/* 下面是 MP3 语音模块的驱动控制函数，通过调用这些函数实现语音模块的控制 */
/**/
#include "reg24le1.h"
/**

```c
/*函数名称：sendout()
/*函数功能：利用串口发送一个字符
/*输入参数：dat 为待发送的字符
/*返回参数：无
/*******************************************************************/
void sendout(unsigned char dat)
{
    S0BUF = dat;
    while(!TI0);                   /*等待数据传送*/
    TI0 = 0;
}
/*******************************************************************
/*函数名称：sendmp3()
/*函数功能：MP3 模块控制命令发送函数
/*输入参数：com 为支持的控制命令，turn 和 dat 参数根据 com 不同而起不同作用
/*返回参数：无
/*******************************************************************/
void sendmp3(unsigned char com,unsigned char dat,int turn)
{
 unsigned char t;
 switch(com)
  {
    case 0xa4 :{sendout(0x7e);sendout(0x03);sendout(0xa4);sendout(dat);sendout(0x7e);}break;
    case 0xa0 :{sendout(0x7e);              /*MP3 播放命令*/
                sendout(0x07);
                sendout(0xa0);
                t = dat/10 + 0x30;
                sendout(t);
                t = dat%10 + 0x30;
                sendout(t);
                t = turn/100 + 0x30;
                sendout(t);
                t = (turn%100)/10 + 0x30;
                sendout(t);
                t = turn%10 + 0x30;
                sendout(t);
                sendout(0x7e);
                }break;
    case 0xa1 :{sendout(0x7e);sendout(0x02);sendout(0xa1);sendout(0x7e);}break;
```

```
        case 0xa2 :{sendout(0x7e);sendout(0x02);sendout(0xa2);sendout(0x7e);}break;
        case 0xa3 :{sendout(0x7e);sendout(0x02);sendout(0xb3);sendout(0x7e);}break;
        case 0xc0 :{sendout(0x7e);sendout(0x02);sendout(0xc0);sendout(0x7e);}break;
        case 0xc1 :{sendout(0x7e);sendout(0x04);sendout(0xc1);sendout(dat/10 + 0x30);sendout(dat %
                   10 + 0x30);sendout(0x7e);}break;
        case 0xc2 :{sendout(0x7e);sendout(0x03);sendout(0xc2);sendout(0xaa);sendout(0x7e);}break;
    }
}
/******************************************************************
/函数名称：MP3_play()
/函数功能：MP3 播放控制函数
/输入参数：x 为播放曲目
/返回参数：无
******************************************************************/
void MP3_play(unsigned char x)
{
  sendmp3(0xa0,0x01,x);
}
/******************************************************************
/函数名称：Vol_con()
/函数功能：MP3 音量控制
/输入参数：x 为设置的音乐播放音量
/返回参数：无
******************************************************************/
void Vol_con(unsigned char x)
{
  sendmp3(0xa4,x,0);
}
/******************************************************************
/函数名称：MP3_stop()
/函数功能：MP3 模块放音暂停
/输入参数：无
/返回参数：无
******************************************************************/
void MP3_stop(void)
{
  sendmp3(0xa1,0,0);
}
```

第4章　nRF24LE1 的射频收发器与应用

```c
/************************************************************
/函数名称:MP3_replay()
/函数功能:MP3 模块恢复放音
/输入参数:无
/返回参数:无
************************************************************/
void MP3_replay(void)
{
sendmp3(0xa2,0,0);
}
/************************************************************
/函数名称:MP3_rst()
/函数功能:实现 MP3 模块软件复位
/输入参数:无
/返回参数:无
************************************************************/
void MP3_rst(void)
{
sendmp3(0xc2,0,0);
}
/************************************************************
/                语音模块的控制函数部分到此结束             /
************************************************************/
```

3. nRF24LE1 无线收发功能配置程序

nRF24LE1 无线收发功能部分相关的程序定义如下:

```c
/************************************************************
/宏定义一些简单的函数,简化函数
************************************************************/
# include "reg24le1.h"
# include "wireless_api.h"
# include "rf24le1.h"
/************************************************************/
# define CSN_LOW()      RFCSN = 0;
# define CSN_HIGH()     RFCSN = 1;
# define CE_LOW()       RFCE = 0;Timer_10Us(1);
# define CE_HIGH()      RFCE = 1;
# define CE_PULSE()     CE_HIGH();Timer_10Us(2);CE_LOW();
```

```c
/***************************************************
/用到的全局变量的定义
****************************************************/
idata unsigned char radio_status;                    /*无线状态*/
static unsigned char gnDataPipeNumber;               /*无线通信的数据通道号*/
code const address[] = {0x12,37,0x55,0x79,0x97};     /*无线收发地址设置*/
/***************************************************
/函数名称: Rf24L01_Reg_RW()
/函数功能: nRF24LE1 的无线部分 SPI 接口读/写操作
/输入参数: byte 为 SPI 接口要传输的一个数据
/返回参数: 之前的状态值
****************************************************/
static unsigned char Rf24L01_Reg_RW(unsigned char byte)
{
    SPIRDAT = byte;                 /*写入要传送的一字节*/
    RFSPIF = 0;                     /*清除标志*/
    while(!RFSPIF);                 /*等待完成*/
    return SPIRDAT;                 /*返回状态*/
}
/***************************************************
/函数名称: Rf24L01_WriteByte()
/函数功能: 向指定无线配置寄存器中写入指定的值
/输入参数: reg 为寄存器地址, value 为要写入的值
/返回参数: reg 中原来的值
****************************************************/
unsigned char Rf24L01_WriteByte(unsigned char reg, unsigned char value)
{
    unsigned char status;
    CSN_LOW();
    status = Rf24L01_Reg_RW(reg);   /*选定要写入的寄存器*/
    Rf24L01_Reg_RW(value);          /*写入值*/
    CSN_HIGH();
    return(status);                 /*返回寄存器状态*/
}

/***************************************************
/函数名称: Rf24L01_WriteMultiByte()
/函数功能: 指定的寄存器, 并一次性连续写入多个值
/输入参数: reg 为寄存器地址, pBuf 为一个数据指针, len 为写入数据的个数
```

/返回参数: 起始值
**/

```c
unsigned char Rf24L01_WriteMultiByte(unsigned char reg, unsigned char * pBuf, unsigned char len)
{
    unsigned char status,i;
    CSN_LOW();
    status = Rf24L01_Reg_RW(reg);              /*选定要写的寄存器*/
    for(i = 0; i<len; i++)                     /*将要写的值写入缓冲区*/
    {
        Rf24L01_Reg_RW(*pBuf++);
    }
    CSN_HIGH();
    return(status);                            /*返回状态位*/
}
```

/**
/函数名称: Rf24L01_Clear_IRQ()
/函数功能: 清除无线中断请求标志
/输入参数: irq_flag 为用于清除中断的标志
/返回参数: 传输状态值
**/

```c
unsigned char Rf24L01_Clear_IRQ(unsigned char irq_flag)
{
    return Rf24L01_WriteByte(WRITE_REG + STATUS, irq_flag);
}
```

/**
/函数名称: Rf24L01_Flush_TX()
/函数功能: 刷新无线发送缓冲区
/输入参数: 无
/返回参数: 无
**/

```c
void Rf24L01_Flush_TX(void)
{
    Rf24L01_WriteByte(FLUSH_TX,0);
}
```

/**
/函数名称: Rf24L01_Flush_RX()
/函数功能: 刷新无线接收缓冲区
/输入参数: 无
/返回参数: 无

```
*****************************************************************/
void Rf24L01_Flush_RX(void)
{
    Rf24L01_WriteByte(FLUSH_RX,0);
}
/*****************************************************************
/函数名称：Rf24L01_ReadByte()
/函数功能：读取一个寄存器的值
/输入参数：reg 为待读的寄存器地址
/返回参数：读取的寄存器的值
*****************************************************************/
unsigned char Rf24L01_ReadByte(unsigned char reg)
{
    unsigned char reg_val;
    CSN_LOW();
    Rf24L01_Reg_RW(reg);                  /*选定要读的寄存器*/
    reg_val = Rf24L01_Reg_RW(0);          /*读取寄存器值*/
    CSN_HIGH();
    return(reg_val);                      /*返回寄存器值*/
}
/*****************************************************************
/函数名称：Rf24L01_RxTx_Switch()
/函数功能：实现无线的发送和接收模式的切换
/输入参数：bMode 为待设定的无线的工作模式
/返回参数：无
*****************************************************************/
void Rf24L01_RxTx_Switch(unsigned char bMode)
{
    unsigned char bConfig;
    CE_LOW();
    Rf24L01_Flush_RX();                   /*清空收发缓冲区*/
    Rf24L01_Flush_TX();
    Rf24L01_Clear_IRQ(MASK_IRQ_FLAGS);
    bConfig = Rf24L01_ReadByte(CONFIG);   /*读取当前无线收发用配置寄存器的值*/
    if(bMode == PRX)
    {
        if((bConfig&0x01))
        {
            CE_HIGH();
```

```c
            return;
        }
        bConfig &= 0xfe;
        bConfig |= 0x01;
        Rf24L01_WriteByte(WRITE_REG + CONFIG, bConfig);
        CE_HIGH();
    }
    else if(bMode == PTX)
    {
        if(!(bConfig&0x01))
        {
            return;
        }
        bConfig &= 0xfe;
        Rf24L01_WriteByte(WRITE_REG + CONFIG, bConfig);
    }
}
/******************************************************************
/函数名称：Rf24L01_ReadMultiByte()
/函数功能：从一个寄存器地址连续读出 len 个数据，存入 pBuf 指向的存储单元
/输入参数：reg 为寄存器地址，pBuf 为指向存储单元的指针，len 为数据个数
/返回参数：状态量
******************************************************************/
static unsigned char Rf24L01_ReadMultiByte(unsigned char reg, unsigned char * pBuf, unsigned char len)
{
    unsigned char status,i;
    CSN_LOW();
    status = Rf24L01_Reg_RW(reg);              /* 选定操作的寄存器 */
    for (i = 0;i<len;i++)
    {
        pBuf[i] = Rf24L01_Reg_RW(0);           /* 读取数据 */
    }
    pBuf[i] = '\0';
    CSN_HIGH();
    return(status);                            /* 返回状态字 */
}
/******************************************************************
/函数名称：Rf24L01_RX()
```

/函数功能：用于接收有效的无线数据
/输入参数：prx_bug 为指向数据存储单元的指针
/返回参数：无
***/
```c
static void Rf24L01_RX(unsigned char * prx_buf)
{
    unsigned char gnRF_RecvLen;
    gnRF_RecvLen = Rf24L01_ReadByte(READ_PAYLAODLEN);      /*读取接收数据长度*/
    if (prx_buf != NULL)
        Rf24L01_ReadMultiByte(RD_RX_PLOAD,prx_buf,gnRF_RecvLen); /*读取全部数据*/
}
```
/***
/函数名称：Rf24L01_TX()
/函数功能：利用无线发送一个字符串
/输入参数：ptx_buf 为指向待发送的数据单元的指针，nLen 为待发送的数据个数
/返回参数：无
***/
```c
void Rf24L01_TX(unsigned char * ptx_buf,unsigned char nLen)
{
    Rf24L01_RxTx_Switch(PTX);                   /*转换到发送模式*/
    Rf24L01_Flush_TX();                         /*清空发送缓冲区*/
    Rf24L01_WriteMultiByte(WR_TX_PLOAD, ptx_buf, nLen); /*连续写入要发送的数据*/
    CE_PULSE();                                 /*20 μs 的高脉冲,激发数据发送*/
}
```
/***
/函数名称：Rf24L01_TXABYTE()
/函数功能：nRF24LE1 无线发送一字节的数据
/输入参数：x 为待发送的一字节的数据
/返回参数：无
***/
```c
void Rf24L01_TXABYTE(unsigned char x)
{
    Rf24L01_RxTx_Switch(PTX);                   /*转换到发送模式*/
    Rf24L01_Flush_TX();                         /*清除发送缓冲区*/
    Rf24L01_WriteByte(WR_TX_PLOAD, x);          /*写入要发送的数据*/
    CE_PULSE();                                 /*20 μs 的高脉冲*/
}
```
/***
/中断服务处理函数，用于处理无线传输中的一些情况

```c
*******************************************************************/
static void rf_rdy_ov_interrupt(void) interrupt INTERRUPT_RFRDY {}
static void rfirq_ov_interrupt(void) interrupt INTERRUPT_RFIRQ
{
    radio_status = 0xFF;
}
/*******************************************************************
/函数名称：Rf24L01_Polling_IRQ()
/函数功能：通过扫描中断标志的方式接收无线发来的数据
/输入参数：rev_buf 为指向数据存储单元的指针
/返回参数：当前数据状态
*******************************************************************/
unsigned char Rf24L01_Polling_IRQ(unsigned char * rev_buf)
{
    unsigned char irq_status;
    if(radio_status == 0x00)
    {
        if(Rf24L01_ReadByte(FIFO_STATUS) & MASK_RX_EMPTY) return IDLE;
        Rf24L01_RX(rev_buf);                            /*将收到的有效数据存到 revbuf 指针当中*/
        irq_status = Rf24L01_ReadByte(STATUS);          /*读取 SPI 状态寄存器*/
        gnDataPipeNumber = (irq_status >> 1)&0x7;       /*取当前收到数据的通道号*/
        return (unsigned char)RX_DR;                    /*成功地接收到数据*/
    }
    radio_status = 0x00;
    irq_status = Rf24L01_ReadByte(STATUS);              /*读无线收发用状态寄存器*/
    switch(irq_status&MASK_IRQ_FLAGS)                   /*取 4、5、6 位*/
    {
        case MASK_RX_DR_FLAG:                           /*接收到数据*/
        {
            if(Rf24L01_ReadByte(FIFO_STATUS) & MASK_RX_EMPTY)  /*接收缓冲区空*/
            {
                irq_status = IDLE;
            }
            else
            {
                Rf24L01_RX(rev_buf);
                gnDataPipeNumber = (irq_status >> 1)&0x7;
                irq_status = (unsigned char)RX_DR;
            }
```

```c
                break;
            }
            case MASK_TX_DS_FLAG:                    /*带自动应答的数据发送成功*/
            {
                irq_status = (unsigned char)TX_DS;
                break;
            }
            case MASK_MAX_RT_FLAG:                   /*最大重发触发了中断*/
            {
                irq_status = (unsigned char)MAX_RT;
                break;
            }
            case IDLE:                               /*没有数据*/
            {
                irq_status = (unsigned char)IDLE;
                break;
            }
    }
    if(Rf24L01_ReadByte(FIFO_STATUS) & MASK_RX_EMPTY)   /*接收空,清除标志*/
        Rf24L01_Clear_IRQ(MASK_IRQ_FLAGS);
        return irq_status;
}
/************************************************************
/函数名称: Rf24L01_Set_Init()
/函数功能: 初始化nRF24LE1无线收发功能
/输入参数: drc为指向SetupData类型的一个无线配置结构体的指针
/返回参数: 一个配置是否完成的一个标志量
*************************************************************/
unsigned char Rf24L01_Set_Init(SetupData * drc)
{
    unsigned char btemp;
    RFCKEN = 1;                              /*使能无线传输时钟*/
    RF = 0;                                  /*RF中断清零*/
    CE_LOW();                                /*选中RF*/
    Rf24L01_Flush_RX();                      /*清空接收发送缓冲*/
    Rf24L01_Flush_TX();
    Rf24L01_Clear_IRQ(MASK_IRQ_FLAGS);       /*清除中断请求状态寄存器*/
    /**********************************************************
    /设置无线的配置寄存器,这个很重要
```

```c
                  ******************************************/
btemp = 0x0e;
if (drc ->nOn == Dev_Normal) btemp += 0x01;
Rf24L01_WriteByte(WRITE_REG + CONFIG, btemp);
/*******************************************************
/设置通道自动应答
  ******************************************/
Rf24L01_WriteByte(WRITE_REG + EN_AA, drc ->nAutoAck);
/*******************************************************
/使能接收通道地址配置
  ******************************************/
Rf24L01_WriteByte(WRITE_REG + EN_RXADDR, drc ->nRecvAddr);
/*******************************************************
/设置无线模块地址宽度
  ******************************************/
Rf24L01_WriteByte(WRITE_REG + SETUP_AW, 0x03);
/*******************************************************
/设置重传延时和最多重传的次数
  ******************************************/
btemp = (drc ->nART_Factor&0x0f) << 4;
btemp += (drc ->nRetran > 15 ? 15 : drc ->nRetran);
btemp = 0x17;//self
Rf24L01_WriteByte(WRITE_REG + SETUP_RETR, btemp);
/*******************************************************
/设置无线收发的工作频点
  ******************************************/
if (drc ->nChannel>122)
{
    drc ->nChannel = 0;
}
Rf24L01_WriteByte(WRITE_REG + RF_CH, drc ->nChannel);
/*******************************************************
/设置无线收发的传输速率和传输功率
  ******************************************/
btemp = ((drc ->nPower&0x0f)>3 ? 3:(drc ->nPower&0x0f)) << 1;
btemp += 0x01;
if((drc ->nPower&0xf0)>0) btemp += 0x08;
btemp = 0x0e;
Rf24L01_WriteByte(WRITE_REG + RF_SETUP, btemp);
```

```c
/**************************************************************
/设置通道 0 地址和数据宽度
**************************************************************/
Rf24L01_WriteMultiByte(WRITE_REG + RX_ADDR_P0,drc->aAddr1,5);
if (drc->nLen>32)
{
    drc->nLen = 32;
}
else if (drc->nLen == 0)
{
    drc->nLen = 1;
}
Rf24L01_WriteByte(WRITE_REG + RX_PW_P0, drc->nLen);
btemp = drc->nRecvAddr;
btemp = btemp >> 1;
if (btemp&1)
{
/**************************************************************
/设置接收通道 1 的地址
**************************************************************/
    Rf24L01_WriteMultiByte(WRITE_REG + RX_ADDR_P1,drc->aAddr2,5);
/**************************************************************
/设置接收通道 1 的数据宽度
**************************************************************/
    Rf24L01_WriteByte(WRITE_REG + RX_PW_P1, drc->nLen);
}
/**************************************************************
/设置无线收发的地址
**************************************************************/
Rf24L01_WriteMultiByte(WRITE_REG + TX_ADDR,drc->aAddr1,5);
/**************************************************************
/设置无线收发用的特征寄存器
**************************************************************/
Rf24L01_WriteByte(WRITE_REG + FEATURE, 0x05);
/**************************************************************
/设置无线收发的动态数据长度
**************************************************************/
Rf24L01_WriteByte(WRITE_REG + DYNPD, drc->nRecvAddr);
RF = 1;
```

```c
        if(drc->nOn == Dev_Normal)
        {
            CE_HIGH();
        }
        return 0;
}
/***************************************************************/
/函数名称: wireless_init()
/函数功能: nRF24LE1 的无线部分的总初始化程序
/输入参数: 无
/返回参数: 无
****************************************************************/
void wireless_init(void)
{
    SetupData mSetup;
    memset(&mSetup,0,sizeof(SetupData));        /*将结构体 SetupData 初始化为 0*/
    mSetup.nChannel = 20;
    mSetup.nPower = 3;                           /*功率设置成 0 dB*/
    mSetup.nRecvAddr = 0x01;
    mSetup.nAutoAck = 0x01;
    mSetup.nLen = 32;
    mSetup.nRetran = 8;
    mSetup.nOn = 1;
    memcpy(mSetup.aAddr1,address,5);             /*address 地址复制到前面一个空间*/
    mSetup.nOn = 1;
    Rf24L01_Set_Init((SetupData *)&mSetup);     /*调用结构体初始化无线部分*/
}
/***************************************************************
/                    无线功能程序到此结束                      /
****************************************************************/
```

4. nRF24LE1 单片机控制程序

nRF24LE1 单片机的控制程序如下:

```c
/***************************************************************
/nRF24LE1 内部集成了增强型 51 单片机内核,下面是一些相关资源的初始化
****************************************************************/
#include "reg24le1.h"
```

```c
#include "config.h"
#include "intrins.h"
#include "Sys.h"
/**************************************************************/
volatile unsigned char Udata;
char minute = 0, second = 0;
/**************************************************************
/函数名称: system_init()
/函数功能: nRF24LE1 的系统初始化
/输入参数: 无
/返回参数: 无
***************************************************************/
void system_init(void)
{
    EA = 0;
    CLKCTRL = 0x28;              /* 设置 nRF24LE1 的工作主时钟 */
    CLKLFCTRL = 0x01;            /* 设置 nRF24LE1 实时时钟源 */
    Timer_Ms(50);
    SPIRCON1 = 0x0B;
    RFCKEN = 1;                  /* 使能 nRF24LE1 内部无线时钟 */
    RF = 0;
    WUIRQ = 0;                   /* 唤醒中断使能 */
    PODIR &= 0x7F;
    P07 = 0;
}
/**************************************************************
/函数名称: time1_init()
/函数功能: 初始化定时器 1
/输入参数: 无
/返回参数: 无
***************************************************************/
void time1_init(void)
{
TMOD = 0x10;
TH1 = (65536 - 50000)/256;
TL1 = (65536 - 50000) % 256;
ET1 = 1;
EA = 1;
TR1 = 1;
```

}
/**
/函数名称：Timer_Us()
/函数功能：软件延时 10 μs
/输入参数：无
/返回参数：无
**/
```c
void Timer_Us(void)
{
    _nop_();_nop_();_nop_();_nop_();
    _nop_();_nop_();_nop_();_nop_();
    _nop_();_nop_();_nop_();_nop_();
    _nop_();_nop_();_nop_();_nop_();
    _nop_();_nop_();_nop_();
}
```
/**
/函数名称：Timer_10Us
/函数功能：软件延时指定的 10 μs 数
/输入参数：nTime 为延时的 10 μs 数
/返回参数：无
**/
```c
void Timer_10Us(unsigned char nTime)
{
    while((nTime--)!=0) Timer_Us();
}
```
/**
/函数名称：Timer_Ms()
/函数功能：软件延时指定的毫秒数
/输入参数：nTime 为延时的 ms 数
/返回参数：无
**/
```c
void Timer_Ms(int nTime)
{
    int j;
    while((nTime--)!=0)
    for(j=0;j<350;j++);
}
```
/**
/函数名称：timer1svr()

```
/函数功能：处理定时器 1 中断
/输入参数：无
/返回参数：无
***************************************************************/
void timer1svr() interrupt INTERRUPT_TF1
{
    static char flag = 0;
    TR1 = 0;
    TH1 = (65536 - 50000)/256;
    TL1 = (65536 - 50000)%256;
    flag++;
    if(flag == 20)
    {
        flag = 0;
        second++;
        P07 = !P07;
        if(second == 60)
        {
            second = 0;
            minute++;
            if(minute == 60)
            minute = 0;
        }
    }
    TR1 = 1;
}
/****************************************************************
/函数名称：putchar()
/函数功能：利用 nRF24LE1 串口发送一个字符
/输入参数：dat 为待发送的字符
/返回参数：无
***************************************************************/
void putchar( unsigned char dat)
{
    S0BUF = dat;
    while(!TI0);
    TI0 = 0;
}
/****************************************************************
```

```
/函数名称：puts()
/函数功能：串口发送一个字符串
/输入参数：s为指向要发送的字符串
/返回参数：无
******************************************************************/
void ( char *s )
{
    while( *s != '\0')
    {
    putchar(*s);
    s++;
    }
}
/*****************************************************************
/                    单片机控制程序到此结束                       /
******************************************************************/
```

5. 常用函数

常用函数如下：

```
/*包含基本的延时函数、串口调试函数以及按键扫描函数的定义*/
/******************************************************************/
#include "reg24le1.h"
#include "config.h"
/******************************************************************
/函数名称：delay()
/函数功能：软件延时函数
/输入参数：x为软件延时的时间数
/返回参数：无
******************************************************************/
void delay(unsigned int x)
{
 int j;
 for(;x>0;x--)
    for(j=200;j>0;j--)
        {
        ;
        }
}
```

```
/*****************************************************************
/函数名称:debug()
/函数功能:串口调试函数,通过串口可以显示相关的数据
/输入参数:ptr 为指向调试显示的内容
/返回参数:无
*****************************************************************/
void debug(char * ptr)
{
        putchar(ptr[0] + '0');
        putchar(ptr[1] + '0');
        putchar(ptr[2]/10 + '0');
        putchar(ptr[2] % 10 + '0');
        putchar(ptr[3] + '0');
        putchar(ptr[4] + '0');
        putchar('\n');
}


/*****************************************************************
/函数名称:getkey()
/函数功能:扫描键盘
/输入参数:ps 为指向一个存储区的指针
/返回参数:键盘按下状态
*****************************************************************/
unsigned char getkey(char * ps)
{
 P1DIR = 0xFF;
 P1 = 0xFF;
 P1CON = 0xD0;                    /*扫描按键1*/
 if(P10 == 0)
 {
 delay(5);
 if(P10 == 0)
 {
 while(!P10);
 return next;
 }
 }
 P1CON = 0xD1;                    /*扫描按键2*/
 if(P11 == 0)
```

```
            {
            delay(5);
            if(P11 == 0)
            {
            while(!P11);
            return back;
            }
            }
        P1CON = 0xD2;                      /*扫描按键3*/
        if(P12 == 0)
        {
            delay(5);
            if(P12 == 0)
            {
            while(!P12);
            return volp;
            }
        }
        P1CON = 0xD3;                      /*扫描按键4*/
        if(P13 == 0)
        {
            delay(5);
            if(P13 == 0)
            {
            while(!P13);
            return vold;
            }
        }

        P1CON = 0xD4;                      /*扫描按键5*/
        if(P14 == 0)
        {
            delay(5);
            if(P14 == 0)
            {
            while(!P14);
            if(ps[0])
            return play;
            else if(ps[1])
```

```
    return stop;
  }
 }
 return 0;

}
/**************************************************************
/                     以上函数定义到此结束                      /
**************************************************************/
```

6. 主函数

本程序的主函数部分如下：

```
/**************************************************************
/通过在主函数中调用前面定义的各个子功能函数,可以实现各个相关的功能
**************************************************************/
#include "reg24le1.h"
#include "config.h"
#include "wireless_api.h"
#include "rf24le1.h"
#include "LCD.h"
#include "Sys.h"
#include "mp3.h"
/**************************************************************
/宏定义
**************************************************************/
#define SEND_24LE1              /*如果屏蔽,就是接收;如果有该宏定义,就是发射*/
#define max    100
#define min      0
#ifdef SEND_24LE1
char buffer[6] = {1,0,1,4,0,0};    /*存储定义*/
#else
char rbuffer[6] = {0,0,0,4,0,0};   /*存储接收到的数据*/
#endif
unsigned char xuhao = 0;
void main(void)
{
    char num = 0;
    int step = 0;
```

第4章 nRF24LE1 的射频收发器与应用

```
    system_init();                  /*nRF24LE1 控制器部分的初始化*/
    wireless_init();                /*nRF24LE1 无线模块的初始化*/
    #ifdef SEND_24LE1               /*如果定义了 SEND_24LE1,就是发送端*/
    baudinint(38400);               /*串口波特率设置成 38 400 用来控制液晶*/
    time1_init();                   /*定时器 1 初始化*/
    #else
    baudinint(9600);                /*波特率设置成 9 600*/
    delay(20);
    MP3_rst();
    #endif

    #ifdef SEND_24LE1
    delay(100);
    DwinClear();                    /*液晶清除显示*/
    DwinPicture(19);                /*设置启动界面*/
    while(step<200)                 /*在启动时刻显示精度条*/
    {
    DwinJingdu(50,100,step);
    step++;
    delay(100);
    }
    delay(2000);
    DwinClear();                    /*液晶显示器清屏*/
    DwinPicture(20);                /*显示存储在液晶显示器里面的第 20 个位置的图片*/
    dispword();
    #endif
                                    /*根据宏定义 SEND_24LE1 设置无线的工作模式是收还是发*/
    #ifdef SEND_24LE1
    Rf24L01_RxTx_Switch(PTX);
    #else
    Rf24L01_RxTx_Switch(PRX);
    #endif
    delay(10000);                   /*延时 10 000*/
    #ifdef SEND_24LE1
    DwinPicture(17);
    #endif
    EA = 1;
    while(1)
    {
```

第4章 nRF24LE1 的射频收发器与应用

```c
#ifdef SEND_24LE1                              /*无线发送端*/
showvol(buffer[3]);
showtitle(buffer[2],minute,second);            /*显示标题*/
num = getkey(buffer);                          /*输入的状态*/
switch(num)
{
/************************************************************
/说明部分：buffer[0]是暂停标志位,buffer[1]是播放标志位,buffer[2]存储播放的音乐编
         号,buffer[3]是当前音量号,buffer[4]是静音
 ************************************************************/
case stop:buffer[0] = 1;buffer[1] = 0;showstop();TR0 = 0;break;
case play:buffer[0] = 0;buffer[1] = 1;showplay();TR0 = 1;break;
case next:buffer[2] ++ ;if(buffer[2]>max)buffer[2] = min;shownext();break;
case back:buffer[2] -- ;if(buffer[2]<min)buffer[2] = max;showback();break;
case volp:buffer[3] ++ ;if(buffer[3]>7)buffer[3] = 8;break;
case vold:buffer[3] -- ;if(buffer[3]<1)buffer[3] = 0;break;
case voln:if(buffer[4])buffer[4] = 0;else buffer[4] = 1;break;
default:break;
}
if(num!= 0)                                    /*如果有按键按下*/
{
Rf24L01_TX((unsigned char * )buffer,6);        /*发送控制命令*/
delay(5);
debug(buffer);
P07 = !P07;
}
#else                                          /*无线接收端*/
if (Rf24L01_Polling_IRQ((unsigned char * )rbuffer) == RX_DR)  /*收到数据*/
{
P07 = !P07;
if(rbuffer[0] == 1)
{
MP3_stop();                                    /*MP3 播放暂停*/
}
else
{
 MP3_replay();                                 /*MP3 恢复播放*/
}
Vol_con(rbuffer[3]);
```

第4章 nRF24LE1 的射频收发器与应用

```
            if(xuhao != rbuffer[2])           /* 如果当前指定的曲目和正在播放的
                                                 不一致,就播放当前的指定曲目 */
            {
               MP3_play(rbuffer[2]);
               xuhao = rbuffer[2];
            }
            if(rbuffer[4])                    /* 如果静音标志置位,则设置音量为最低 */
            Vol_con(0);
            }
          #endif
        }
    }
/*****************************************************************
                     本程序所有代码到此结束
*****************************************************************/
```

第 5 章

nRF24LE1 与常用外围模块的连接及编程

5.1 nRF24LE1 与数码管和键盘的连接及编程

5.1.1 nRF24LE1 与 ZLG7289 的连接

ZLG7289B 是广州周立功单片机发展有限公司自行设计的数码管显示驱动和键盘扫描管理芯片,可直接驱动 8 位共阴式数码管(或 64 只独立 LED),同时还可以扫描管理多达 64 只按键。ZLG7289B 内部含有显示译码器,可直接接收 BCD 码或十六进制码,并同时具有 2 种译码方式。此外,还具有消隐、闪烁、左移、右移、段寻址等多种控制指令。ZLG7289B 采用 SPI 串行总线与微控制器接口,仅占用少数几根 I/O 口线。利用片选信号,多片 ZLG7289B 还可以并接在一起使用,能够方便地实现多于 8 位的显示或多于 64 只按键的应用。

ZLG7289B 可广泛地应用于仪器仪表、工业控制器、条形显示器及控制面板等领域。

ZLG7289B 与数码管和键盘连接的电路如图 5.1.1 所示。

nRF24LE1 与 ZLG7289B 的电路连接关系如表 5.1.1 所列。

表 5.1.1 nRF24LE1 与 ZLG7289B 的连接

nRF24LE1 的 I/O	ZLG7289B 的接口
P0.0	CS
P0.2	CLK
P0.7	DIO
P0.6	INT

5.1.2 nRF24LE1 与 ZLG7289 的编程示例

本程序运行初始化后,8 位数码管显示 0~7 的 8 个数字。这时如果按下 S0,全部数码管就会显示 0;按下 S1,全部数码管就会显示 1,以此类推。如果按下按键 S10 和 S11,那么就会显示一些乱码,因为数码管是设置在十进制显示模式下,只能显示 0~9 十个数字。程序运行

第 5 章 nRF24LE1 与常用外围模块的连接及编程

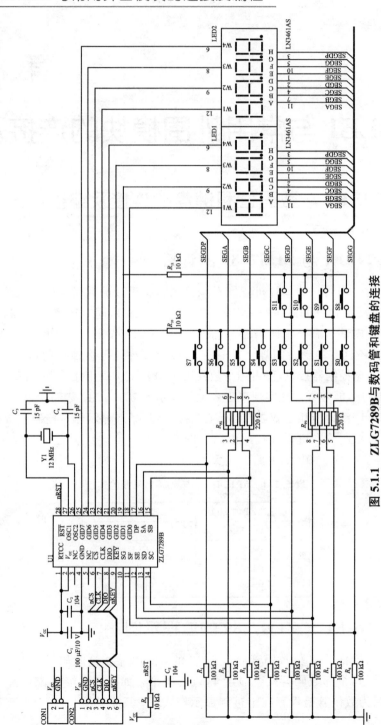

图 5.1.1 ZLG7289B 与数码管和键盘的连接

时,也可通过串口将每个按下的按键键值发送到计算机,显示在串口调试软件的接收窗口。

1. 程序定义

nRF24LE1 与 ZLG7289 程序相关定义如下:

```
/******************************************************/
#include "reg24le1.h"
/******************************************************/
/nRF24LE1 与 ZLG7289 键盘管理和数码显示芯片
/******************************************************/
#define    DIO          P07
#define    SCLK         P02
#define    INT          P06
#define    CS           P00
#define    Disableint   do{EA = 0;}while(0)
#define    Enableint    do{EA = 1;}while(0)
#define    BAUD_2400    2400
#define    BAUD_4800    4800
#define    BAUD_9600    9600
#define    BAUD_14400   14400
#define    BAUD_19200   19200
#define    BAUD_38400   38400
typedef    unsigned int    uint;
typedef    unsigned char   uchar;
/******************************************************
/函数名称:Delay()
/函数功能:实现软件延时
/输入参数:x 为延时的时间数
/返回参数:无
/******************************************************/
void Delay(uint x)
{
uchar l;
    for(;x>0;x--)
      for(l = 12;l>0;l--)
        {}
}
/******************************************************
/函数名称:SetMcuClk()
/函数功能:nRF24LE1 工作时钟设置
```

```
/输入参数:无
/返回参数:无
***************************************************************/
void SetMcuClk()
{
    CLKCTRL = 0x28;                    /* 使用 XCOSC 16 MHz */
    CLKLFCTRL = 0x01;
}
```

2. nRF24LE1 串口功能控制程序

```
/***************************************************************
/函数名称:Uartinit()
/函数功能:nRF24LE1 的串口初始化
/输入参数:baud 为设置的波特率
/返回参数:无
***************************************************************/
void Uartinit(uint baud)
{
    P0DIR &= 0xF7;                     /* 配置 P0.3(TXD)为输出 */
    P0DIR |= 0x10;                     /* 配置 P0.4(RXD)为输入 */
    S0CON = 0x50;                      /* 串口模式选择允许接收 */
    PCON |= 0x80;                      /* 波特率倍增 */
    WDCON |= 0x80;                     /* 选用内部波特率发生器 */
    switch(baud)
    {
        case 38400:
            {
                S0RELL = 0xF3;
                S0RELH = 0x03;
            }
            break;
        case 19200:
            {
                S0RELL = 0xE6;
                S0RELH = 0x03;
            }
            break;
        case 14400:
            {
```

```
                    S0RELL = 0xDE;
                    S0RELH = 0x03;
                }
                break;
        case 9600:
                {
                    S0RELL = 0xCC;
                    S0RELH = 0x03;
                }
                break;
        case 4800:
                {
                    S0RELL = 0x98;
                    S0RELH = 0x03;
                }
                break;
        case 2400:
                {
                    S0RELL = 0x30;
                    S0RELH = 0x03;
                }
                break;
        default:
                {
                    S0RELL = 0xCC;
                    S0RELH = 0x03;
                }
                break;
        }
    return ;
}
/*****************************************************************
/函数名称：putch()
/函数功能：串口发送一个字符
/输入参数：ch为待发送的字符
/返回参数：无
*****************************************************************/
void putch(uchar ch)
{
```

```
    S0BUF = ch;
    while(!TI0);
    TI0 = 0;
}
/****************************************************************
/函数名称：puts()
/函数功能：通过 nRF24LE1 串口发送一个字符串
/输入参数：str 为指向一个待发送的字符串的指针
/返回参数：无
*****************************************************************/
void puts(char * str)
{
    while( * str != '\0')
    {
        putch( * str ++ );
    }
    Delay(10);
}
```

3. ZLG7289 的初始化函数和接口函数

```
/****************************************************************
/函数名称：ZLG7289init()
/函数功能：ZLG7289 的初始化
/输入参数：无
/返回参数：无
*****************************************************************/
void ZLG7289init()
{
    P0DIR = (BIT_6|BIT_7);       /* 初始化 INT1 和 DIO 为输入 */
    CS = 1;
    SCLK = 1;
    DIO = 1;
    INT = 1;
    INTEXP |= BIT_4;             /* INT1 使能 */
    IT0 = 1;                     /* 下降沿触发 */
    EX0 = 1;                     /* 开启外部中断 */
}
/****************************************************************
/函数名称：WriteSPI()
```

```
/函数功能：用 nRF24LE1 的 I/O 口模拟 SPI 接口写一字节函数
/输入参数：Dat 为模拟 SPI 要传输的字节
/返回参数：无
******************************************************/
void WriteSPI(uchar Dat)
{
    uchar cT = 8;
    P0DIR & = (~(BIT_7));
    do {                               /*循环写一字节的数据*/
        if((Dat & 0x80) == 0x80) {
            DIO = 1;
        } else {
            DIO = 0;
        }
        Dat ≪ = 1;
        SCLK = 1;
        Delay(5);
        SCLK = 0;
        Delay(5);
    } while ( -- cT != 0);
}
/******************************************************
/函数名称：ReadSPI()
/函数功能：通过模拟的 SPI 接口读取一字节数据
/输入参数：无
/返回参数：从模拟 SPI 接口读取到的一字节
******************************************************/
uchar ReadSPI(void)
{
    uchar cDat = 0;
    uchar cT = 8;
    P0DIR |= BIT_7;
    do {                               /*循环读一字节的数据*/
        SCLK = 1;
        Delay(5);
        cDat ≪ = 1;
        if (DIO) {
            cDat ++ ;
        }
```

```
            SCLK = 0;
            Delay(5);
      } while ( -- cT != 0);
      return cDat;
}
/***************************************************************
/函数名称: WriteCmd()
/函数功能: 通过模拟的 SPI 接口给 ZLG7289B 写命令字
/输入参数: Cmd 为命令字
/返回参数: 无
****************************************************************/
void WriteCmd(uchar Cmd)
{
   CS = 0;
   Delay(25);
   WriteSPI(Cmd);
   CS = 1;
   Delay(5);
}
/***************************************************************
/函数名称: WriteCmdDat()
/函数功能: 通过模拟的 SPI 接口给 ZLG7289B 写指令和数据
/输入参数: Cmd 为命令字,Dat 为数据
/返回参数: 无
****************************************************************/
void WriteCmdDat(uchar Cmd,uchar Dat)
{
   CS = 0;
   Delay(25);
   WriteSPI(Cmd);
   Delay(15);
   WriteSPI(Dat);
   CS = 1;
   Delay(5);
}
/***************************************************************
/函数名称: ZLG7289DownLoad()
/函数功能: 在数码管上指定的位置显示指定的字符
/输入参数: Mod 为 ZLG7289B 的译码方式,Cx 为数码管显示的位
```

cDp 决定是否显示小数点,cDat 为要在数码管上显示的数
/返回参数:无
***/

```
void ZLG7289DownLoad(uchar Mod,uchar cX,uchar cDp,uchar cDat)
{
    uchar ucModDat[3] = {0x80,0xC8,0x90};
    uchar ucD1;
    uchar ucD2;
    if (Mod > 2)
        {
         Mod = 2;
        }
    ucD1 = ucModDat[Mod];
    cX &= 0x07;
    ucD1 |= cX;
    ucD2 = cDat & 0x7F;
    if (cDp == 1) {
        ucD2 |= 0x80;
    }
    WriteCmdDat(ucD1, ucD2);
}
```

/**
/函数名称:GetKey()
/函数功能:读取键盘按下的按键键值
/输入参数:无
/返回参数:按键键值
**/

```
uchar GetKey()
{
    uchar cKey;
    CS = 0;
    Delay(25);
    WriteSPI(0x15);
    Delay(15);
    cKey = ReadSPI();
    CS = 1;
    Delay(5);
    return cKey;
}
```

4. 外部中断服务函数

```
/***************************************************************
/函数名称：EXTISR()
/函数功能：处理外部中断，主要是读取按键键值和显示按键键值
/输入参数：无
/返回参数：无
***************************************************************/
uchar KeyValue = 0;
void EXTISR() interrupt INTERRUPT_IFP
{
    EX0 = 0;
    KeyValue = GetKey();
    ZLG7289DownLoad (0, 0, 0, KeyValue);
    ZLG7289DownLoad (0, 1, 0, KeyValue);
    ZLG7289DownLoad (0, 2, 0, KeyValue);
    ZLG7289DownLoad (0, 3, 0, KeyValue);
    ZLG7289DownLoad (0, 4, 0, KeyValue);
    ZLG7289DownLoad (0, 5, 0, KeyValue);
    ZLG7289DownLoad (0, 6, 0, KeyValue);
    ZLG7289DownLoad (0, 7, 0, KeyValue);
    putch('0' + (KeyValue/10));
    putch('0' + (KeyValue%10));
    putch('\n');
    EX0 = 1;
}
```

5. 主函数

```
/***************************************************************
/主函数部分
***************************************************************/
void main()
{
    /*初始化*/
    Disableint;
    SetMcuClk();
    Uartinit(BAUD_9600);
    ZLG7289init();
```

```
    zlg7289Reset();
    EnableInt;
    Delay(100);
    ZLG7289DownLoad (0, 0, 0, 0);
    ZLG7289DownLoad (0, 1, 0, 1);
    ZLG7289DownLoad (0, 2, 0, 2);
    ZLG7289DownLoad (0, 3, 0, 3);
    ZLG7289DownLoad (0, 4, 0, 4);
    ZLG7289DownLoad (0, 5, 0, 5);
    ZLG7289DownLoad (0, 6, 0, 6);
    ZLG7289DownLoad (0, 7, 0, 7);
    while(1);
}
/***************************************************************
/                       程序到此结束                             /
****************************************************************/
```

5.2 nRF24LE1 与液晶显示器模块的连接及编程

5.2.1 RT12864M 汉字图形点阵液晶显示器模块简介

RT12864M 汉字图形点阵液晶显示器模块可显示汉字及图形,内置 8 192 个中文汉字(16×16 点阵)、128 个字符(8×16 点阵)及 64×256 点阵显示 RAM(GDRAM)。LCD 类型为 STN,显示内容 128 列×64 行,显示颜色为黄绿色,电源电压范围为 3.3~5 V(内置升压电路,无需负压),与 MCU 接口为 8 位或 4 位并行,或者 3 位串行,配置 LED 背光,外观尺寸为 93 mm×70 mm×12.5 mm,视域尺寸为 73 mm×39 mm。

RT12864M 液晶显示器模块的引脚端功能如表 5.2.1 所列。

表 5.2.1 RT12864M 液晶显示器模块的引脚端功能

引脚	符号	功能
1	V_{SS}	模块的电源地
2	V_{DD}	模块的电源正端
3	V_0	LCD 驱动电压输入端
4	RS(CS)	并行的指令/数据选择信号;串行的片选信号
5	R/W(SID)	并行的读/写选择信号;串行的数据口
6	E(CLK)	并行的使能信号;串行的同步时钟
7	DB0	数据 0

第 5 章 nRF24LE1 与常用外围模块的连接及编程

续表 5.2.1

引脚	符号	功能
8	DB1	数据 1
9	DB2	数据 2
10	DB3	数据 3
11	DB4	数据 4
12	DB5	数据 5
13	DB6	数据 6
14	DB7	数据 7
15	PSB	并/串行接口选择：高电平为并行；低电平为串行
16	NC	空引脚
17	$\overline{RST}$	复位，低电平有效
18	OUT	LCD 驱动电压输出端
19	A	背光源正极(LED+5 V)
20	K	背光源负极(LED−0 V)

RT12864M 液晶显示器模块有并行和串行两种连接方法。RT12864M 液晶显示器模块基本指令集和扩充指令集。更多的内容请参考"RT12864M 液晶显示器模块数据手册"。

5.2.2 nRF24LE1 与 RT12864M 的连接

nRF24LE1 与 RT12864M 液晶显示器之间采用串行方式连接，这样可以节省 nRF24LE1 的 I/O 口，同时连接也方便。液晶显示器模块的接口电路如图 5.2.1 所示。

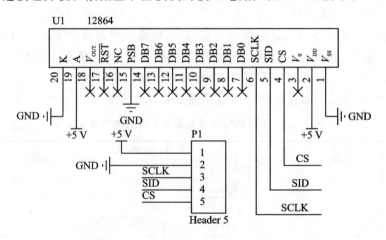

图 5.2.1　液晶显示器模块的接口电路

nRF24LE1 的 I/O 口与液晶显示器模块的连接如表 5.2.2 所列。

第5章 nRF24LE1 与常用外围模块的连接及编程

表 5.2.2　nRF24LE1 的 I/O 口与 12864 液晶显示器模块的连接

nRF24LE1 的 I/O 口	12864 液晶显示器模块
P0.0	CS
P0.1	SID
P0.2	SCLK

5.2.3　nRF24LE1 与液晶显示器模块的编程示例

在本示例程序中,利用 nRF24LE1 作为控制器,在液晶显示器模块上显示一屏汉字。本示例程序的流程图如图 5.2.2 所示。

程序代码如下：

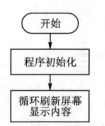

图 5.2.2　示例程序流程图

1. 程序定义

```
/*****************************************************
/ nRF24LE1 驱动 12864 液晶显示程序
*****************************************************/
#include "reg24le1.h"
#include "intrins.h"
/*****************************************************/
sbit    CS = P0^0;
sbit    E_CLK = P0^2;
sbit    RW_SID = P0^1;
/*****************************************************/
#define Enableint   do{EA = 1;}while(0)
#define Disableint  do{EA = 1;}while(0)
/*****************************************************/
typedef enum   YON{false,true}bool;
/*****************************************************/
typedef unsigned char uchar;
typedef unsigned int  uint ;
/*****************************************************/
uchar code tab[] = {
"这是一个液晶测试"
"小江编写,仅供简"
"程序,由南华大学"
"单测试,谢谢支持"
};
```

2. nRF24LE1 的初始化功能函数

```
/******************************************************
/函数名称：IOCNFG()
/函数功能：nRF24LE1 的 I/O 口初始化
/输入参数：无
/返回参数：无
******************************************************/
void IOCNFG()
{
    P0DIR& = 0xF0;    /*设置 P0.0~P0.3 为输出*/
}
/******************************************************
/函数名称：WorkClkSet()
/函数功能：nRF24LE1 工作时钟设置
/输入参数：无
/返回参数：无
******************************************************/
void WorkClkSet()
{
    CLKCTRL = 0x28;
    CLKLFCTRL = 0x01;
}
/******************************************************
/函数名称：delay()
/函数功能：实现延时
/输入参数：n 为延时的时间数
/返回参数：无
******************************************************/
void delay(uint n)
{
    uint i;
    for(i = 0; i<n; i++);
}
```

3. 液晶显示器模块的串口数据传送

```
/******************************************************
/函数名称：SendByte()
/函数功能：串行发送一字节数据
```

/输入参数：dat 为模拟 SPI 接口给液晶显示器发送的一字节数据
/返回参数：无
***/
```c
void SendByte(uchar dat)
{
 unsigned   char   i;
 CS = 1;
 for(i = 0;i<8;i++)
 {
 E_CLK = 0;
 if(dat&0x80)
 {
 RW_SID = 1;
 }
 else
 {
 RW_SID = 0;
 }
 _nop_();
 _nop_();
 E_CLK = 1;
 dat = dat << 1;
 _nop_();
 _nop_();
 }
 CS = 0;
}
```

/***
/函数名称：SendCMD()
/函数功能：通过模拟的 SPI 接口给液晶显示器发送控制命令
/输入参数：dat 为给液晶显示器发送的指令
/返回参数：无
***/
```c
void SendCMD(uchar dat)
{
 SendByte(0xF8);
 SendByte(dat&0xF0);
 SendByte((dat&0x0F) << 4);
```

第5章 nRF24LE1 与常用外围模块的连接及编程

```
}
/******************************************************************
/函数名称：SendDat()
/函数功能：发送一字节的数据
/输入参数：dat 为给液晶显示器发送的数据
/返回参数：无
******************************************************************/
void SendDat(uchar dat)
{
    SendByte(0xFA);
    SendByte(dat&0xF0);
    SendByte((dat&0x0F) << 4);
}
```

4. 液晶显示器模块的初始化函数

```
/******************************************************************
/函数名称：initlcm()
/函数功能：LCD 初始化函数
/输入参数：无
/返回参数：无
******************************************************************/
void initlcm(void)
{
    delay(100);
    SendCMD(0x30);    /* 功能设置，一次送 8 位数据，基本指令集 */
    SendCMD(0x0C);    /* 开显示 */
    SendCMD(0x01);    /* 清 DDRAM */
    SendCMD(0x02);    /* DDRAM 地址归位 */
    SendCMD(0x80);    /* 设定 DDRAM */
    delay(100);
}
```

5. 液晶显示器模块的汉字显示函数

```
/******************************************************************
/函数名称：chn_disp()
/函数功能：显示汉字或字符
/输入参数：chn 为指向要显示的字符串的指针
/返回参数：无
```

```
**************************************************/
void chn_disp (uchar code * chn)
{
    uchar i,j;
    SendCMD(0x80);
    for (j = 0;j<4;j++)
     {
      for (i = 0;i<16;i++)
      {
        SendDat(chn[j * 16 + i]);
      }
     }
}
```

6. 主函数

```
/**************************************************
/主函数
**************************************************/
void main(void)
{
    WorkClkSet();
    IOCNFG();
    initlcm();              /*12864初始化程序*/
    while(1)
    {
    chn_disp(tab);
    delay(1000);
    }
}
/**************************************************
/                程序代码到此结束                    /
**************************************************/
```

5.3 nRF24LE1 与 DAC 的连接及编程

5.3.1 nRF24LE1 与 DAC TLC5615 的连接

TLC5615 是带有缓冲基准输入(高阻抗)的 10 位电压输出 DAC(数/模转换器)。DAC 具有基准电压两倍的输出电压范围,而且是单调变化的。器件使用 5 V 单电源工作。器件具有上电复位功能,以确保可重复启动。

TLC5615 通过 3 线串行总线控制,接口与 CMOS 的微控制器接口兼容。可以采用 SPI、QSPI、Microwire 标准数字通信协议。

TLC5615 采用 8 引脚的小型封装,允许在空间受限制地使用。TLC5615C 的工作温度范围为 0 ℃～70 ℃。TLC5615I 的工作温度范围为 -40 ℃～85 ℃。

采用 TLC5615 构成的数/模转换电路如图 5.3.1 所示。

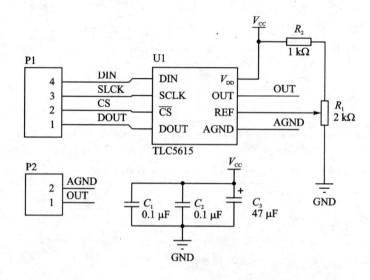

图 5.3.1　TLC5615 数/模转换电路

nRF24LE1 的 I/O 口与 TLC5615 模块的电路连接关系如表 5.3.1 所列。

表 5.3.1　nRF24LE1 的 I/O 口与 TLC5615 模块的连接

nRF24LE1 的 I/O 口	TLC5615 模块
P0.0	SCLK
P0.1	DIN
P0.2	$\overline{CS}$

5.3.2　nRF24LE1 与 DAC 的编程示例

本示例程序将演示使用 TLC5615 产生正弦波输出。程序中采用的查表法,根据时间来输出一个与正弦波形对应的电压量。如果希望使输出的正弦波的波形更加完美,则需要增加正弦表的表值量。本示例程序的流程图 5.3.2 所示,程序源代码如下所示。

第 5 章　nRF24LE1 与常用外围模块的连接及编程

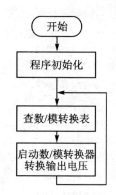

图 5.3.2　示例程序的流程图

1. 程序定义

```
#include "reg24le1.h"
#include "intrins.h"
/*******************************************************/
#define   SPI_CLK      P00
#define   SPI_DATA     P01
#define   CS_DA        P02
#define   LED          P03
#define   Disableint   do{EA = 0;}while(0)
#define   Enableint    do{EA = 1;}while(0)
/*******************************************************/
typedef unsigned int uint;
typedef unsigned char uchar;
/*******************************************************/
```

2. 正弦波的码表定义

```
code uint DA_data[256] = {              /*生成正弦波的码表*/
0x1FF,0x20C,0x218,0x225,0x231,0x23E,0x24A,0x256,
0x263,0x26F,0x27B,0x287,0x293,0x29F,0x2AB,0x2B7,
0x2C3,0x2CE,0x2DA,0x2E5,0x2F0,0x2FB,0x306,0x311,
0x31B,0x326,0x330,0x33A,0x343,0x34D,0x357,0x360,
0x369,0x371,0x37A,0x382,0x38A,0x392,0x39A,0x3A1,
0x3A8,0x3AF,0x3B6,0x3BC,0x3C2,0x3C8,0x3CD,0x3D3,
0x3D8,0x3DC,0x3E1,0x3E5,0x3E8,0x3EC,0x3EF,0x3F2,
0x3F5,0x3F7,0x3F9,0x3FB,0x3FC,0x3FD,0x3FE,0x3FE,
0x3FE,0x3FE,0x3FE,0x3FD,0x3FC,0x3FB,0x3F9,0x3F7,
```

0x3F5,0x3F2,0x3EF,0x3EC,0x3E8,0x3E5,0x3E1,0x3DC,
0x3D8,0x3D3,0x3CD,0x3C8,0x3C2,0x3BC,0x3B6,0x3AF,
0x3A8,0x3A1,0x39A,0x392,0x38A,0x382,0x37A,0x371,
0x369,0x360,0x357,0x34D,0x343,0x33A,0x330,0x326,
0x31B,0x311,0x306,0x2FB,0x2F0,0x2E5,0x2DA,0x2CE,
0x2C3,0x2B7,0x2AB,0x29F,0x293,0x287,0x27B,0x26F,
0x263,0x256,0x24A,0x23E,0x231,0x225,0x218,0x20C,
0x1FF,0x1F2,0x1E6,0x1D9,0x1CD,0x1C0,0x1B4,0x1A8,
0x19B,0x18F,0x183,0x177,0x16B,0x15F,0x153,0x147,
0x13B,0x130,0x124,0x119,0x10E,0x103,0x0F8,0x0ED,
0x0E3,0x0D8,0x0CE,0x0C4,0x0BB,0x0B1,0x0A7,0x09E,
0x095,0x08D,0x084,0x07C,0x074,0x06C,0x064,0x05D,
0x056,0x04F,0x048,0x042,0x03C,0x036,0x031,0x02B,
0x026,0x022,0x01D,0x019,0x016,0x012,0x00F,0x00C,
0x009,0x007,0x005,0x003,0x002,0x001,0x000,0x000,
0x000,0x000,0x000,0x001,0x002,0x003,0x005,0x007,
0x009,0x00C,0x00F,0x012,0x016,0x019,0x01D,0x022,
0x026,0x02B,0x031,0x036,0x03C,0x042,0x048,0x04F,
0x056,0x05D,0x064,0x06C,0x074,0x07C,0x084,0x08D,
0x095,0x09E,0x0A8,0x0B1,0x0BB,0x0C4,0x0CE,0x0D8,
0x0E3,0x0ED,0x0F8,0x103,0x10E,0x119,0x124,0x130,
0x13B,0x147,0x153,0x15F,0x16B,0x177,0x183,0x18F,
0x19B,0x1A8,0x1B4,0x1C0,0x1CD,0x1D9,0x1E6,0x1F2
                };

3. TLC5615 接口函数

```
/****************************************************************
/函数名称：DA_TLC_5615()
/函数功能：用模拟 SPI 接口给 DAC 转换器 TLC5615 发送数据，让其进行转换输出
/输入参数：Dat 为待进行数/模转换的数字量
/返回参数：无
****************************************************************/
void DA_TLC_5615(uint Dat)
{
    uchar i;
    Dat <<= 6;
    SPI_CLK = 0;              /*时钟低*/
    CS_DA = 0;                /*片选有效*/
    for (i = 0;i<12;i++)
```

```
    {
        if((Dat&0x8000) == 0)
        {
            SPI_DATA = 0;
        }
        else
        {
            SPI_DATA = 1;
        }
        SPI_CLK = 1;              /*时钟高*/
        Dat <<= 1;                /*左移一位*/
        SPI_CLK = 0;              /*时钟低*/
    }
    CS_DA = 1;
}
```

4. nRF24LE1 初始化

```
/**************************************************************
/函数名称：WorkClkSet()
/函数功能：nRF24LE1 工作时钟设置函数
/输入参数：无
/返回参数：无
**************************************************************/
void WorkClkSet()
{
    CLKCTRL = 0x28;
    CLKLFCTRL = 0x01;
}
/**************************************************************
/函数名称：IOCNFG()
/函数功能：初始化 nRF24LE1 的 I/O 口
/输入参数：无
/返回参数；无
**************************************************************/
void IOCNFG()
{
    P0DIR &= 0xF0;              /*设置控制器的 P0.0～P0.3 为输出*/
    P0DIR = 0x00;
    CS_DA = 1;                  /*DAC 芯片不被选择*/
}
```

第 5 章 nRF24LE1 与常用外围模块的连接及编程

5. LED 灯闪烁函数

```
/**************************************************************
/函数名称：Light_LED()
/函数功能：使 LED 按照一定的时间间隔闪烁
/输入参数：无
/返回参数：无
**************************************************************/
void Light_LED()
{
  static uint Cnt = 0;
  if(Cnt != 10000)
    {
      Cnt ++ ;
    }
  else
    {
      Cnt = 0;
      LED = !LED;
    }
}
```

6. 主函数

```
/**************************************************************
/主函数
**************************************************************/
void main(void)
{
  uint k;
  Disableint;
  WorkClkSet();
  IOCNFG();
  Enableint;
  while(1)
  {
    k ++ ;
    if(k> = 256)
    {
      k = 0;
```

```
        }
        DA_TLC_5615(DA_data[k]);
        Light_LED();
    }
}
/***************************************************************/
/                         程序结束                              /
/***************************************************************/
```

5.4 nRF24LE1 与 DDS 的连接及编程

5.4.1 nRF24LE1 与 DDS AD9850 的连接

1. DDS AD9850 简介

AD9850 是一个集成有高速数字频率合成器 DDS、高性能高速度数/模转换器和比较器的器件,可以构成一个直接可编程的频率合成器和时钟发生器。当采用一个精确的时钟脉冲信号源时,AD9850 可以产生一个稳定的频率和相位可编程的数字化的模拟正弦波输出。这个正弦波可以直接作为频率源,或在芯片内部被改变为方波在时钟发生器中应用。AD9850 的高速 DDS 核心采用一个 32 位的频率调谐字,在采用一个 125 MHz 系统时钟情况下,具有大约为 0.029 1 Hz 的输出调谐分辩率。AD9850 提供有关 5 位的相位调制分辩率,能够移相 180°、90°、45°、22.5°、11.25°。

AD9850 的频率调谐字、控制器字和相位调制字通过并行或串行的装载格式异步装入 AD9850。并联装载格式由 5 个重复装入的 8 位控制字组成。第 1 个 8 位字节控制相位调制,激活低功耗和装载格式;剩余的 2~5 字节组成 32 位频率调谐字。串行装载采用 40 位的串行数据流通过并行输入总线中的一个来完成。

AD9850 采用了先进的 CMOS 技术,电源电压为 3.3 V 或 5 V,功耗为 380 mW @ 125 MHz (5 V),功耗为 155 mW @ 110 MHz(3.3 V);低功耗模式的功耗为 30 mW @ 5 V,功耗为 10 mW @ 3.3 V。

有关 AD9850 更多的资料请登录 http://www.analog.com 查询。

采用 AD9850 构成的信号发生电路如图 5.4.1 所示。

2. nRF24LE1 与 AD9850 的连接

采用 nRF24LE1 为控制器件,通过对 AD9850 输送频率控制字,使 AD9850 输出相应频率的正弦波信号。nRF24LE1 与 AD9850 的接口既可采用并行方式,也可采用串行方式。本设计选择串行方式将 nRF24LE1 的 I/O 口接至 AD9850 的串行输入控制端。AD9850 外接

第 5 章　nRF24LE1 与常用外围模块的连接及编程

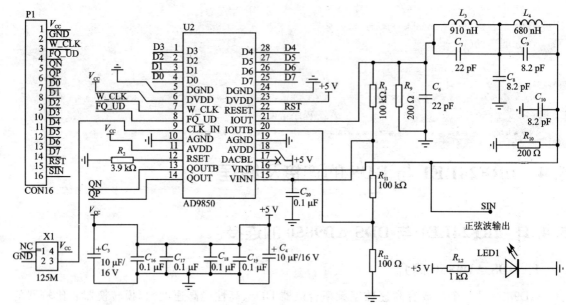

图 5.4.1　AD9850 电路图

125 MHz 的有源晶振,产生的正弦信号经低通滤波器(LPF)去掉高频谐波后,即可得到波形良好的正弦波信号。这样,将 D/A 转换器的输出信号经低通滤波后,接到 AD9850 内部的高速比较器上,即可直接输出一个抖动很小的方波。另外,也可通过键盘编辑任意波形的输出信号。nRF24LE1 的 I/O 口与 AD9850 模块的连接如表 5.4.1 所列。

表 5.4.1　nRF24LE1 的 I/O 口与 AD9850 模块的连接

nRF24LE1 的 I/O 口	AD9850 模块
P0.0	W_CLK
P0.1	FQ_UD
P0.2	RST
P0.3	D7

5.4.2　nRF24LE1 与 DDS 的编程示例

1. 控制程序流程图

本示例程序的流程图如图 5.4.2 所示。

本示例程序控制 AD9850 输出指定频率的正弦波,在示例程序中配置 AD9850 输出频率为 1 MHz 的正弦波。利用高速比较器,可以将输出的正弦波转换成方波,方波信号可以用来作为微控制器的时钟信号。本示例程序源代码如下所示。

第5章 nRF24LE1 与常用外围模块的连接及编程

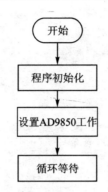

图 5.4.2　nRF24LE1 控制 AD9850 程序流程图

2. 程序定义

```
/****************************************************************
/程序部分开始
****************************************************************/
# include "reg24le1.h"
# include "intrins.h"
/****************************************************************/
# define ad9850_w_clk       P00
# define ad9850_fq_up       P01
# define ad9850_rest        P02
# define ad9850_bit_data    P03
# define LED                P04
# define Enableint          do{EA = 1;}while(0)
# define Disableint         do{EA = 0;}while(0)
# define EnableAD9850       do{ad9850_fq_up = 0;\
                            ad9850_fq_up = 1;}while(0)
```

3. AD9850 操作相关的函数

```
/****************************************************************
/重定义数据类型
****************************************************************/
typedef unsigned char uchar;
typedef unsigned int uint ;
/****************************************************************
/函数名称: ad9850_reset_serial()
/函数功能: AD9850 复位(串口模式)
```

/输入参数:无
/返回参数:无
**/

```c
void ad9850_reset_serial()
{
    ad9850_w_clk = 0;
    ad9850_fq_up = 0;
    ad9850_rest = 0;              /* rest 信号 */
    ad9850_rest = 1;
    ad9850_rest = 0;
    ad9850_w_clk = 0;             /* w_clk 信号 */
    ad9850_w_clk = 1;
    ad9850_w_clk = 0;
    ad9850_fq_up = 0;             /* fq_up 信号 */
    ad9850_fq_up = 1;
    ad9850_fq_up = 0;
}
```

/**
/函数名称:Write_Byte_S()
/函数功能:往 AD9850 写入一字节(串行模式)
/输入参数:Dat 为待写入的数据
/返回参数:无
**/

```c
void Write_Byte_S(uchar Dat)
{
    uchar i;
    for(i = 0;i<8;i++)
    {
        ad9850_bit_data = (Dat >> i)&0x01;
        ad9850_w_clk = 1;
        ad9850_w_clk = 0;
    }
}
```

/**
/函数名称:ad9850_wr_serial()
/函数功能:串口方式写 AD9850 的控制字
/输入参数:w0 为控制字 0,frequence 为设置的输出波形频率
/返回参数:无
**/

```c
void ad9850_wr_serial(uchar w0,double frequence)
{
uchar   w;
long int y;
double   x;
x = 4294967295/125;
frequence = frequence * x/1000000;
y = frequence;
w = (y >> = 0);
Write_Byte_S(w);          /*写 w4 数据*/
w = (y >> 8);
Write_Byte_S(w);          /*写 w3 数据*/
w = (y >> 16);
Write_Byte_S(w);          /*写 w2 数据*/
w = (y >> 24);
Write_Byte_S(w);          /*写 w1 数据*/
w = w0;
Write_Byte_S(w);          /*写 w0 数据*/
EnableAD9850;             /*移入使能*/
}
```

4. nRF24LE1 初始化函数和延时函数

```c
/***************************************************************
/函数名称：WorkClkSet()
/函数功能：nRF24LE1 工作时钟设置
/输入参数：无
/返回参数：无
***************************************************************/
void WorkClkSet()
{
CLKCTRL = 0x28;
CLKLFCTRL = 0x01;
}
/***************************************************************
/函数名称：IO_CNFG()
/函数功能：nRF24LE1 的 I/O 口配置函数，P0 口配置成输出，初始化为低电平
/输入参数：无
/返回参数：无
***************************************************************/
```

```
void IO_CNFG()
{
P0DIR = 0x00;
P0 = 0x00;
}
/***************************************************************
/函数名称：Delay()
/函数功能：实现软件延时
/输入参数：x 为软件延时的时间数
/返回参数：无
***************************************************************/
void Delay(uint x)
{
uint dl;
    for(;x>0;x--)
      for(dl = 120;dl>0;dl--)
        {
          _nop_();
        }
}
/***************************************************************
/函数名称：Light_LED()
/函数功能：控制 LED 灯间隔性地闪烁发光
/输入参数：无
/返回参数：无
***************************************************************/
void Light_LED()
{
LED = !LED;
Delay(1000);
}
```

5. 主函数

```
/***************************************************************
/主函数部分
***************************************************************/
main()
{
 Disableint;
```

```
    WorkClkSet();
    IO_CNFG();
    ad9850_reset_serial();              /*复位芯片 AD9850*/
    Delay(100);
    ad9850_wr_serial(0x00,10000000);    /*串行写入 AD9850*/
    Delay(100);
    Enableint;
    while(1)
    {
        Light_LED();                    /*LED 显示芯片状态*/
    }
}
/******************************************************************
*                         程序到此结束                              *
******************************************************************/
```

5.5 nRF24LE1 与超声波模块的连接及编程

5.5.1 nRF24LE1 与超声波模块的连接

DYP-ME007 超声波测距模块由超声波发射器、接收器和控制电路组成,工作电压为 DC 5 V,工作电流为 15 mA,工作频率为 40 Hz,输入触发信号为 10 μs 的 TTL 脉冲信号,输出信号为 TTL 电平信号,与测量距离成比例,能够实现 3 cm～3.5 m 非接触式的测距功能。

DYP-ME007 的工作原理如下:给予此超声波测距模块一触发信号后,模块发射超声波,当超声波投射到物体而反射回来时,模块输出一回响信号,可以利用触发信号和回响信号间的时间差来判定模块与物体之间的距离。

利用 nRF24L1 的 GPIO 可以控制超声波模块进行测距。DYP-ME007 超声波模块与 nRF24LE1 的连接如图 5.5.1 所示。

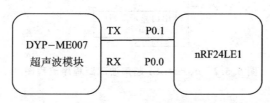

图 5.5.1 超声波模块与 nRF24LE1 的 I/O 接口连接

第5章 nRF24LE1 与常用外围模块的连接及编程

测距过程如下：首先 nRF24LE1 给超声波模块的 TX 端一个大于 10 μs 的高电平触发信号，触发超声波模块发出 8 个 40 kHz 的超声波；然后超声波模块会自动检测是否有信号返回，当检测到超声波信号返回后，RX 端会输出一个高电平信号，高电平信号的持续时间就是超声波从发射到返回的时间。

用上述的超声波模块测距时，只要 nRF24LE1 的 I/O 口控制 TX 端发一个 10 μs 以上的高电平，就可以在接收口(RX)等待高电平输出。当 RX 端从低电平变到高电平时，打开 nRF24LE1 的定时器进行计时，当 RX 端变为低电平时，读取定时器的计时值，此时间即为超声波从发射到返回的时间，根据公式(5.5.1)可以计算出模块与物体之间的距离。

$$测试距离 = (高电平时间 \times 声速)/2 \tag{5.5.1}$$

超声波模块与 nRF24LE1 的 I/O 口的连接关系表如表 5.5.1 所列。

表 5.5.1　超声波模块与 nRF24LE1 的 I/O 口的连接

nRF24LE1 的 I/O 口	超声波模块接口
P0.1	TX
P0.0	RX

5.5.2　nRF24LE1 与超声波模块的编程示例

本示例程序在 nRF24LE1 上运行，利用超声波模块实现超声波测距功能，并将测到的距离反馈回来显示在计算机的显示器上。程序流程图如图 5.5.2 所示，程序源代码如下所示。

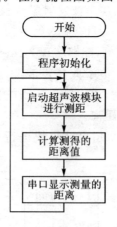

图 5.5.2　nRF24LE1 超声波测距程序流程图

1. 程序定义

```c
/*****************************************************************
/程序说明：nRF24LE1超声波测距,串口显示测量数据
*****************************************************************/
#include "reg24le1.h"
#include   <intrins.h>
/*****************************************************************
/宏定义部分
*****************************************************************/
#define Enableint()   do{EA = 1;}while(0)
#define Disableint()  do{EA = 0;}while(0)
#define true     0x01
#define false    0x00
#define RX       P00
#define TX       P01
#define LED      P02
/*****************************************************************
/类型重新定义
*****************************************************************/
typedef unsigned char uchar;
typedef unsigned int uint;
/*****************************************************************
/定义标志位
*****************************************************************/
bit flag = 0;                    /*溢出标志*/
/*****************************************************************
/函数名称：delayms()
/函数功能：实现软件延时功能
/输入参数：ms为软件延时的毫秒数
/返回参数：无
*****************************************************************/
void delayms(unsigned int ms)
{
    unsigned char i = 100,j;
    for(;ms;ms--)
    {
        while(--i)
        {
            j = 10;
            while(--j);
```

```
    }
  }
}
/***************************************************************
/函数名称：LedDebug()
/函数功能：调用本函数可通过观察 LED 的闪烁来观察程序运行状态
/输入参数：无
/返回参数：无
***************************************************************/
void LedDebug(void)
{
LED = !LED;
delayms(200);
}
/***************************************************************
/函数名称：delayus()
/函数功能：实现软件延时功能
/输入参数：us 为软件延时的微秒数
/返回参数：无
***************************************************************/
void delayus(uint us)
{
us -= 5;
while(us--)
{
_nop_();
}
}
```

2. nRF24LE1 的初始化

```
/***************************************************************
/函数名称：ClkSet()
/函数功能：nRF24LE1 工作时钟的初始化
/输入参数：无
/返回参数：无
***************************************************************/
void ClkSet(void)
{
  CLKCTRL = 0x28;
```

```
    CLKLFCTRL = 0x01;
}
/****************************************************************
/函数名称：IO_init()
/函数功能：初始化 nRF24LE1 的 I/O 口
/输入参数：无
/返回参数：无
****************************************************************/
void IO_init(void)
{
  PODIR |= BIT_0;                /*设置 RX 为输入*/
  PODIR &= (~(BIT_1));           /*设置 TX 为输出*/
  PODIR &= (~(BIT_2));
  RX = 0;
  TX = 0;
}
/****************************************************************
/函数名称：T0_init()
/函数功能：初始化定时器 0
/输入参数：无
/返回参数：无
****************************************************************/
void T0_init(void)
{
  TMOD = 0x01;
  TH0 = 0;
  TL0 = 0;
  ET0 = 1;                       /*允许 T0 中断*/
  TR0 = 0;
}
```

3. nRF24LE1 的串口初始化和功能函数

```
/****************************************************************
/函数名称：Uart_init()
/函数功能：初始化 nRF24LE1 的串口
/输入参数：无
/返回参数：无
****************************************************************/
```

```c
void Uart_init(uint baud)
{
    P0DIR &= 0xF7;              /*配置P0.3(TXD)为输出*/
    P0DIR |= 0x10;              /*配置P0.4(RXD)为输入*/
    P0 |= 0x18;
    S0CON = 0x50;
    PCON |= 0x80;               /*波特率倍增*/
    WDCON |= 0x80;              /*选定内部波特率发生器*/
    if(baud == 38400)
    {
        S0RELL = 0xF3;          /*设置波特率为38 400*/
        S0RELH = 0x03;
    }
    else if(baud == 9600)
    {
        S0RELL = 0xCC;          /*设置波特率为9 600*/
        S0RELH = 0x03;
    }
}
/***************************************************************
/函数名称: putch()
/函数功能: 通过串口输出一个字符
/输入参数: ch为待传输的数据
/返回参数: 无
***************************************************************/
void putch(char ch)
{
    S0BUF = ch;
    while(!TI0);
    TI0 = 0;
}
/***************************************************************
/函数名称: Puts()
/函数功能: 通过串口发送一个字符串
/输入参数: str为一个指向即将被发送的字符串的指针
/返回参数: 无
***************************************************************/
void Puts(char * str)
{
```

```
while( * str != '\0')
 {
 putch( * str ++ );
 }
}
/*****************************************************************
/函数名称:nextline()
/函数功能:让串行口发送的数据显示换行
/输入参数:无
/返回参数:无
*****************************************************************/
void nextline(void)
{
putch('\n');
}
```

4. 超声波模块初始化函数和功能函数

```
/*****************************************************************
/函数名称:StartModule()
/函数功能:启动超声波模块进行距离探测
/输入参数:无
/返回参数:无
*****************************************************************/
void   StartModule()
{
     TX = 1;
     delayus(25);
     TX = 0;
}
/*****************************************************************
/函数名称:Conut()
/函数功能:计算前方障碍物与发射头之间的距离
/输入参数:无
/返回参数:无
*****************************************************************/
void Conut(void)
{
static uint time = 0,TMP;
```

```c
static float S;
time = TH0 * 256 + TL0;
TH0 = 0;
TL0 = 0;
S = (time * 1.87)/100;          /* 算出来是 cm */
if(flag == 1)                    /* 超出测量范围 */
{
flag = 0;
Puts("can,t get the distance!!!");
nextline();
}
else
{
Puts("distance S = ");
TMP = S;
if(TMP >= 100&&TMP<1000)
{
putch((TMP/100) + '0');
putch((TMP%100)/10 + '0');
putch((TMP%10) + '0');
}
else if(TMP >= 10)
{
putch((TMP/10) + '0');
putch((TMP%10) + '0');
}
else if(TMP<10)
{
putch(TMP + '0');
}
putch('.');
TMP = ((int)(S * 100))%100;
putch((TMP/10) + '0');
putch((TMP%10) + '0');
Puts(" cm");
nextline();
}
}
```

5. 中断服务函数

```
/***************************************************************
/函数名称：TIMER0ISR0()
/函数功能：处理定时器中断 0
/输入参数：无
/返回参数：无
***************************************************************/
void TIMER0ISR0() interrupt INTERRUPT_TF0
{
  flag = 1;
}
```

6. 主函数

```
/***************************************************************
/主函数部分
***************************************************************/
void main(void)
{
    static uint timeout = 0;
    Disableint();
    ClkSet();                    /*nRF24LE1 时钟初始化*/
    IO_init();                   /*I/O 口初始化*/
    T0_init();                   /*定时器 0 初始化*/
    Uart_init(38400);            /*波特率初始化*/
    Puts("超声波测距测试程序！");
    nextline();
    Enableint();
    while(1)
    {
     ST : StartModule();         /*启动超声波测距*/
        while(0 == RX)           /*当 RX 为零时等待*/
        {
         delayus(5);
         timeout ++ ;
         if(timeout == 3000)
          {
           timeout = 0;
           goto ST;
          }
```

```
            }
            timeout = 0;
            if(RX == 1)
            TR0 = 1;                    /* 开启计数 */
            while(1 == RX);             /* 当 RX 为 1 时计数并等待 */
            if(RX == 0)
            TR0 = 0;                    /* 关闭计数 */
            Conut();                    /* 计算距离,同时串口显示 */
            LedDebug();                 /* LED 显示程序状态 */
        }
}
/****************************************************************
/                        程序到此结束                             /
****************************************************************/
```

7. 程序运行结果

上面的代码编译通过后,将生成的 HEX 文件下载到 nRF24LE1 上,nRF24LE1 通过串口与计算机连接。复位后程序运行,在计算机的显示器上显示的测量结果如图 5.5.3 所示。改

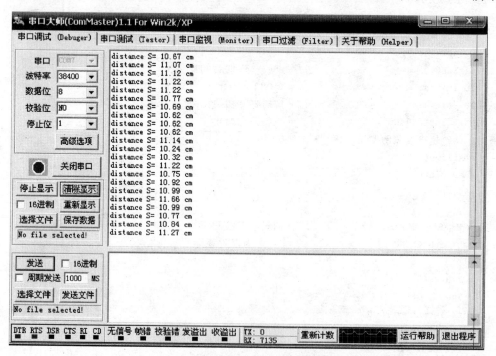

图 5.5.3　超声波测距的测量结果

变超声波模块与前方障碍物间的距离,在显示器上显示的测量值也相应地发生变化。如果测量的距离超出本超声波模块的检测距离,则在显示器上会提示测距无效。

5.6 nRF24LE1 与步进电机驱动模块的连接及编程

5.6.1 nRF24LE1 与步进电机驱动模块的连接

1. 步进电机简介

步进电机可分为反应式步进电机(VR)、永磁式步进电机(PM)和混合式步进电机(HB)。步进电机广泛应用于对精度要求比较高的运动控制系统中,如机器人、打印机、软盘驱动器、绘图仪和机械阀门控制器等。步进电机是数字控制电机,它将脉冲信号转变成角位移,即给一个脉冲信号,步进电机就转动一个角度,即电机的总转动角度由输入脉冲数决定,而电机的转速由脉冲信号频率决定。步进电机的驱动电路根据控制信号工作,控制信号由微控制器产生。

(1) 控制换相顺序

通电换相这一过程称为脉冲分配。例如:混合式步进电机的工作方式,其各相通电顺序为 A→B→C→D,通电控制脉冲必须严格按照这一顺序分别控制 A、B、C、D 相的通断,这就是所谓的脉冲环形分配器。

(2) 控制步进电机的转向

如果给定工作方式正序换相通电,则步进电机正转;如果按反序通电换相,则电机就反转。

(3) 控制步进电机的速度

如果给步进电机发送一个控制脉冲,它就转一步,再发送一个脉冲,它会再转一步。两个脉冲的时间间隔越短,步进电机就转得越快。

2. 步进电机驱动电路

L298N 是专用驱动集成电路,属于 H 桥集成电路,与 L293D 的差别是其输出电流增大,功率增强。其输出电流为 2 A,最高电流为 4 A,最高工作电压为 50 V,可以驱动感性负载,如大功率直流电机、步进电机、减速电机、伺服电机及电磁阀等,特别是其输入端可以与单片机直接相连,从而很方便地受单片机控制。当驱动直流电机时,可以直接控制步进电机,并可以实现电机正转与反转,实现此功能只需改变输入端的逻辑电平。本示例中使用的步进电机驱动芯片是 ST 公司的 L298N,电路如图 5.6.1 所示。

3. nRF24LE1 与步进电机驱动模块的连接

电机驱动模块的网络标号与控制器(nRF24LE1)的 I/O 口的连接关系如表 5.6.1 所列。

第5章 nRF24LE1 与常用外围模块的连接及编程

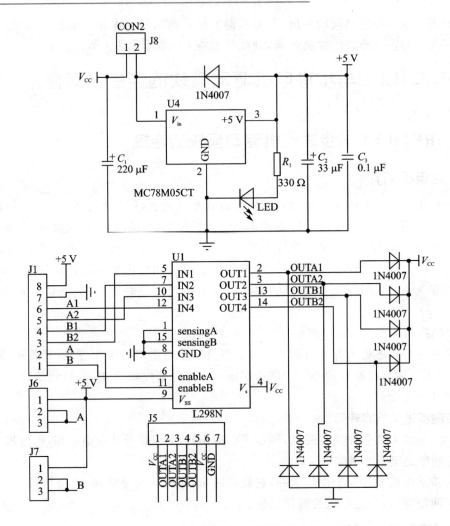

图 5.6.1 步进电机驱动模块电路

表 5.6.1 步进电机驱动模块与 nRF24LE1 的 I/O 口的连接

nRF24LE1 的 I/O 口	步进电机驱动模块网络标号
P0.0	A1
P0.1	A2
P0.2	B1
P0.3	B2

5.6.2　nRF24LE1 与步进电机驱动模块的编程示例

1. 程序流程图

本示例程序的流程图如图 5.6.2 所示。

程序下载运行后,将会控制步进电机驱动模块,使步进电机正转 1 000 步,然后等待一小段时间之后开始反转 1 000 步,这样不断地重复执行这个过程。在程序运行过程中,为了便于观察当前电机的转动方向,设置了两个 LED,分别与 nRF24LE1 的 P0.4 和 P0.5 相连。当电机正转时,与 P0.4 相连的 LED 将会发光。如果步进电机反转,那么与 P0.5 相连的 LED 将会发光。

实现所描述功能的程序代码如下。

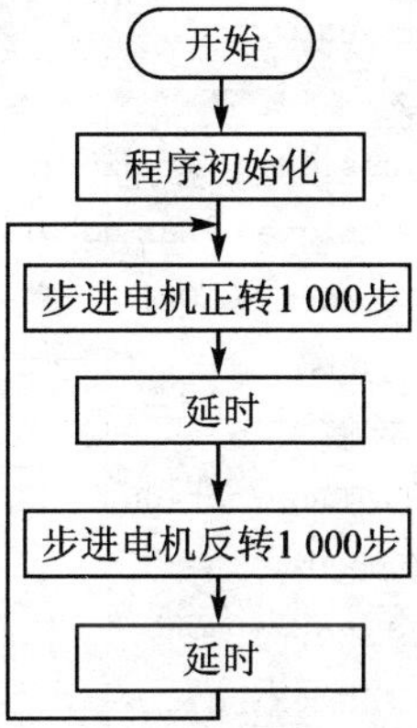

图 5.6.2　nRF24LE1 驱动步进电机程序流程图

2. 程序定义

```
/****************************************************
/nRF24LE1 控制步进电机
****************************************************/
#include "reg24le1.h"
#include "intrins.h"
/****************************************************/
#define  A1    P00
#define  A2    P01
#define  B1    P02
#define  B2    P03
/****************************************************/
#define LEDF P04
#define LEDR P05
#defire Disableint do{EA = 0;}while(0)
#define Enableint  do{EA = 1;}while(0)
#define TABLE_LONGTH    120
/****************************************************/
#define true    1
#define false   0
/****************************************************/
#define Goahead 1
```

```c
#define Goback    0
#define STOP      0
#define RUN       1
/*****************************************************************/
typedef unsigned int uint;
typedef unsigned char uchar;
typedef struct MOTOR{
 uint   Steps;
 uint   Stepc;
 uchar  MOTORderection;
 uint   Speed;
 uchar  State;
}STEPMOTORTYPE;
STEPMOTORTYPE StepMotor;
static   uint xdata   SpeedTable[TABLE_LONGTH];
```

3. nRF24LE1 初始化函数

```c
/******************************************************************
/函数名称：MCUCLKSET()
/函数功能：nRF24LE1 的工作时钟初始化
/输入参数：无
/返回参数：无
*******************************************************************/
void MCUCLKSET()
{
  CLKCTRL = 0x28;            /*使用 XCOSC16M */
  CLKLFCTRL = 0x01;
}
/******************************************************************
/函数名称：IOCONFIG()
/函数功能：初始化 nRF24LE1 的 I/O 口
/输入参数：无
/返回参数：无
*******************************************************************/
void IOCONFIG()
{
  P0DIR &= 0xC0;             /*设置 P0.0～P0.5 为输出 */
  P0 = 0x00;
}
```

```c
/******************************************************************
/函数名称：Timer0init()
/函数功能：初始化定时器0用来控制速度
/输入参数：无
/返回参数：无
******************************************************************/
void Timer0init()
{
TMOD = 0x01;
TH0 = (65536 - 50000)/256;
TL0 = (65536 - 50000)%256;
ET0 = 1;
TR0 = 1;
EA = 0;
}
/******************************************************************
/函数名称：SpeedTableinit()
/函数功能：初始化速度表
/输入参数：无
/返回参数：无
******************************************************************/
void SpeedTableinit()
{
uchar i;
SpeedTable[0] = 30000;
SpeedTable[1] = 25000;
for(i = 2;i<TABLE_LONGTH;i++)
  {
  SpeedTable[i] = SpeedTable[i-2] - SpeedTable[i-1]/4;
  }
}
/******************************************************************
/函数名称：Delay()
/函数功能：软件延时
/输入参数：x为延时的时间数
/返回参数：无
******************************************************************/
void Delay(uint x)
{
```

第 5 章　nRF24LE1 与常用外围模块的连接及编程

```
uchar dl;
   for(;x>0;x--)
    for(dl = 200;dl>0;dl--)
       {
       _nop_();
       }
}
```

4. 步进电机驱动函数

```
/***************************************************************
/函数名称：PulseSet()
/函数功能：设置 4 个控制引脚的电平
/输入参数：_A1、_A2、_B1、_B2 为设置 4 个控制引脚的电平设置值
/返回参数：无
***************************************************************/
void PulseSet(uchar _A1,uchar _A2,uchar _B1,uchar _B2)
{
    A1 = _A1;
    A2 = _A2;
    B1 = _B1;
    B2 = _B2;
    return;
}
/***************************************************************
/函数名称：MOTORRUN()
/函数功能：驱动步进电机转动的函数,调用一次,电机走一步
/输入参数：Derection 为步进电机转动的方向
/返回参数：无
***************************************************************/
void MOTORRUN(uchar Derection)
{
  static uchar Num = 0;
  if(Derection == Goahead)        /*设置正转*/
  {
   Num = (Num + 1) % 8;
  }
  else                            /*设置反转*/
  {
   Num = (Num + 7) % 8;
  }
```

```
  switch(Num)
  {
   case 0 :PulseSet(0,1,1,1);break;
   case 1 :PulseSet(0,0,1,1);break;
   case 2 :PulseSet(1,0,1,1);break;
   case 3 :PulseSet(1,0,0,1);break;
   case 4 :PulseSet(1,1,0,1);break;
   case 5 :PulseSet(1,1,0,0);break;
   case 6 :PulseSet(1,1,1,0);break;
   case 7 :PulseSet(0,1,1,0);break;
   default:break;
  }
}
/*****************************************************************
/函数名称：MotorCtr()
/函数功能：控制步进电机的运行状态和速度
/输入参数：无
/返回参数：无
*****************************************************************/
void MotorCtr()
{
 if(StepMotor.Steps>StepMotor.Stepc)
  {
   StepMotor.State = RUN;
   StepMotor.Speed = StepMotor.Steps - StepMotor.Stepc;
   if(StepMotor.Speed>119)
    {
     StepMotor.Speed = 119;
    }
   MOTORRUN(StepMotor.MOTORderection);
   StepMotor.Stepc ++ ;
  }
 else if(StepMotor.Steps< = StepMotor.Stepc)
  {
   StepMotor.State = STOP;
   StepMotor.Speed = 0;
   StepMotor.Steps = 0;
   StepMotor.Stepc = 0;
  }
}
```

5. 步进电机运行参数设置

```c
/*************************************************************
/函数名称：MOTORSET()
/函数功能：初始化
/输入参数：state 为电机状态，Derection 为转动方向，steps 为设定的步进电机转动步数
/返回参数：无
**************************************************************/
void MOTORSET(uchar state,bit Derection,uint steps)
{
StepMotor.Stepc = 0;
StepMotor.Steps = steps;
StepMotor.MOTORderection = Derection;
StepMotor.Speed = 100;
StepMotor.State = state;
if(StepMotor.State != STOP)
{TR0 = 1;}   /*打开步进电机前进*/
}
```

6. 中断服务函数

```c
/*************************************************************
/函数名称：TIMER0ISR()
/函数功能：处理定时器 0 中断
/输入参数：无
/返回参数：无
**************************************************************/
void TIMER0ISR() interrupt INTERRUPT_TF0
{
 TR0 = 0;
 MotorCtr();
// TH0 = (65536 - SpeedTable[StepMotor.Speed])/256;
// TL0 = (65536 - SpeedTable[StepMotor.Speed]) % 256;
 TH0 = (65536 - 1500)/256;
 TL0 = (65536 - 1500) % 256;
 if(StepMotor.State != STOP)
 {
 TR0 = 1;/*打开定时器*/
 }
}
```

7. 主函数

```
/***********************************************************
/主函数部分
***********************************************************/
void main()
{
/***********************************************************
/初始化函数
***********************************************************/
    Disableint;
    MCUCLKSET();
    IOCONFIG();
    Timer0init();
    SpeedTableinit();
    Enableint;
    while(1)
    {
/***********************************************************
/设置步进电机正转
***********************************************************/
        LEDR = 0;
        LEDF = 1;
        MOTORSET(RUN,Goahead,1000);
        while(StepMotor.State != STOP);
        Delay(20000);
/***********************************************************
/设置步进电机反转
***********************************************************/
        LEDF = 0;
        LEDR = 1;
        MOTORSET(RUN,Goback,1000);
        while(StepMotor.State!= STOP);
        Delay(20000);
    }
}
/***********************************************************
/                        程序到此结束                       /
***********************************************************/
```

第 6 章

Keil μVision4 集成开发环境和 ISP 下载

6.1 Keil μVision4 集成开发环境的使用

Keil C51 是美国 Keil Software 公司出品的 51 系列兼容单片机 C 语言软件开发系统,是目前最流行开发 MCS-51 系列单片机的软件。它提供了包括 C 编译器、宏汇编、连接器、库管理和一个功能强大的仿真调试器等在内的完整开发方案,通过一个集成开发环境(μVision)将这些部分组合在一起。运行 Keil 软件需要 Pentium 或以上的 CPU,16 MB 或更多 RAM、20 MB 以上空闲的硬盘空间,WIN98、NT、WIN2000 和 WINXP 等操作系统。

本节将首先介绍在 Keil μVision4 开发环境里如何建立、编译连接工程及生成工程对应HEX 文件的基本方法。

6.1.1 工程的建立

选择 WINDOWS 操作系统的菜单项"开始"→"程序"→Keil μVision4,启动 Keil μVision4;或双击 Keil μVision4 快捷方式启动。启动 Keil μVision4 如图 6.1.1 所示,启动后的界面如图 6.1.2 所示。

然后选择 Project 菜单项,选择 New μVision4 Project 即弹出 Create New Project 对话框,如图 6.1.3 所示。

接着通过这个对话框选择一个保存工程的目录,最好是英文目录,然后在文件名一栏填上工程的名称,保存类型采用默认值即可,然后会弹出一个 CPU 选型的对话框,如图 6.1.4 所示。

CPU 选型应该选择与当前需要编程的对象一致的 CPU。如果需要编程的对象是nRF24LE1,那么首先需要找到 Nordic Semiconductor,因为 nRF24LE1 是这个公司生产的产品,单击公司名称前面的"+"符号,即可弹出该公司的产品目录,选择 nRF24LE1 即可。

同时右边的小窗口会显示选择的芯片的各种资源配置,CPU 具体选型示例如图 6.1.5 所示。然后单击该窗口上的 OK 按键,接着会弹出一个提示,问是否复制标准的 8051 启动代码添加到刚建立的工程。这需要根据实际情况选择。如果使用的是标准的 8051 内核的单片机,

第 6 章　Keil μVision4 集成开发环境和 ISP 下载

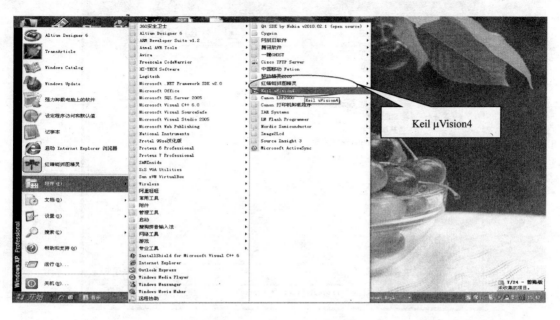

图 6.1.1　Keil μVision4 的启动

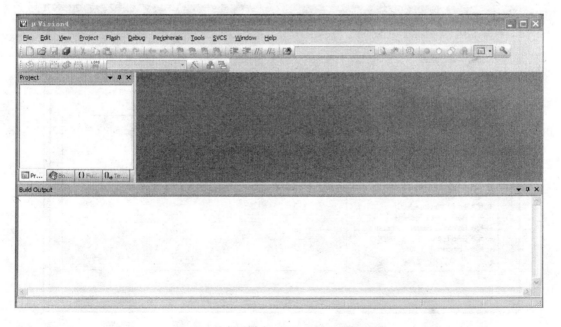

图 6.1.2　Keil μVision4 启动后的主界面

第6章 Keil µVision4 集成开发环境和 ISP 下载

图 6.1.3 Create New Project 对话框

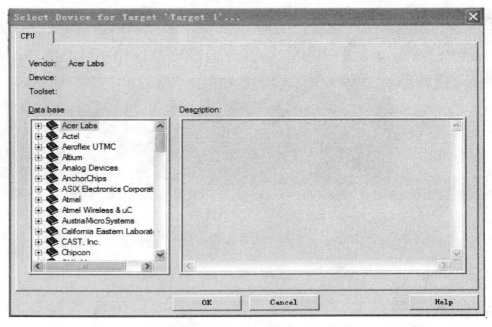

图 6.1.4 CPU 选型对话框

第 6 章 Keil μVision4 集成开发环境和 ISP 下载

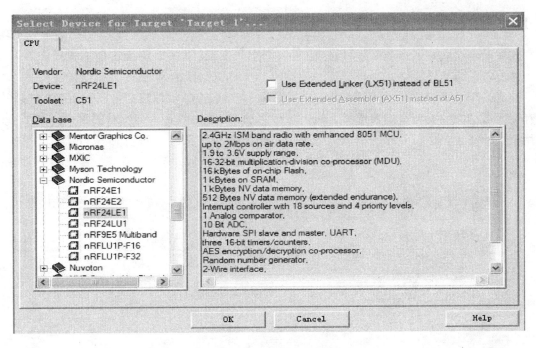

图 6.1.5 CPU 具体选型示例

则选择"是",否则选择"否",这样就完成了一个工程建立。建立好的工程示例如图 6.1.6 所示。

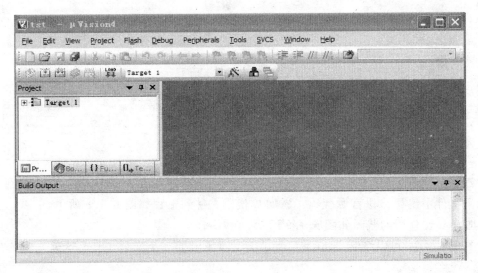

图 6.1.6 刚建立的工程

6.1.2 添加 C 语言文件

现在已经建立好了一个工程,但这还是一个空的工程,下面需要往这个工程里添加 C 语言文件或者汇编文件。

建立一个 C 语言文件的步骤如下:选择 File→New 选项,软件建立好一个文本文件,将它保存为所需的文件类型。接着选择 File→Save 选项,弹出如图 6.1.7 所示的窗口。这里可以采用默认的路径,在文件名一栏填入文件名(推荐选择英文名),文件后缀改为.c 保存,如 tst.c。

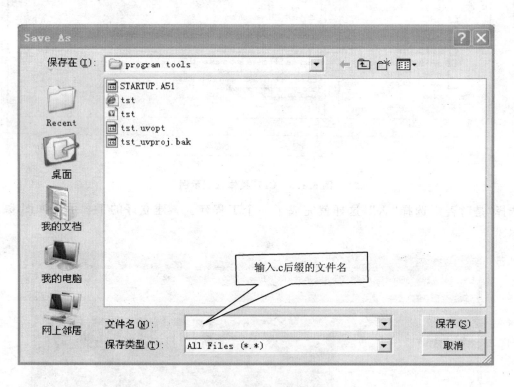

图 6.1.7 建立 C 文件演示

C 文件建立好后,接下来就是将建立好的 C 代码文件添加到工程中,然后在添加到工程中的 C 代码文件中编写 C 语言源代码。添加 C 代码文件到工程如图 6.1.8 所示。

添加好 C 文件后的截图如图 6.1.9 所示。

在图 6.1.9 中,可以看到 Source Group 1 下面多出了一个文件 tst.c,这就是刚添加进去的文件。

第 6 章　Keil μVision4 集成开发环境和 ISP 下载

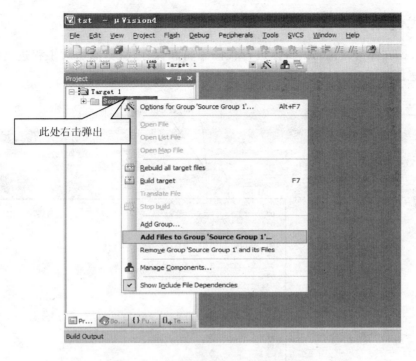

图 6.1.8　将 C 代码文件添加到工程

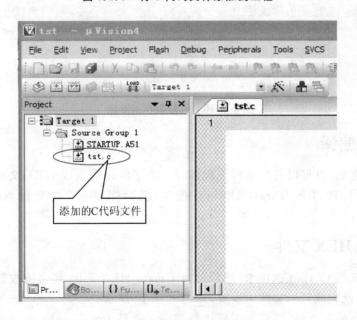

图 6.1.9　添加好 C 代码文件后的工程

6.1.3 代码编辑

双击 tst.c,在弹出的代码编辑框进行代码编辑和保存代码,然后对工程进行编译。一个代码编辑示意图如图 6.1.10 所示。

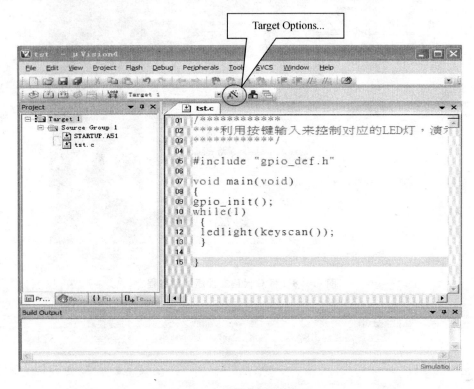

图 6.1.10　代码编辑示意图

6.1.4　工程编译

代码编辑好之后,就可以开始进行工程编译。为了编译后生成 HEX 文件,需要进行输出配置。单击图 6.1.10 中的 Target Options 按钮,就会弹出一个配置设置框,如图 6.1.11 所示。

6.1.5　生成 HEX 文件

单击该设置框的 Output 选项卡,然后勾选 Create HEX File,其余的按默认设置不变,然后单击 OK 按钮,就完成目标文件的设置,如图 6.1.12 所示。

完成上面所有的配置后,即可对工程进行编译。选择 Project→Rebuild all target files 就可以完成工程的编译。如果程序有错误或者警告,都会在 Build Output 窗口中显示出来,根

第 6 章　Keil μVision4 集成开发环境和 ISP 下载

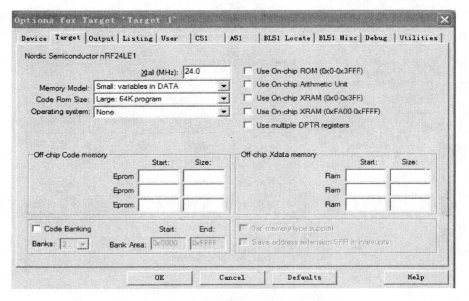

图 6.1.11　目标配置选项设置框

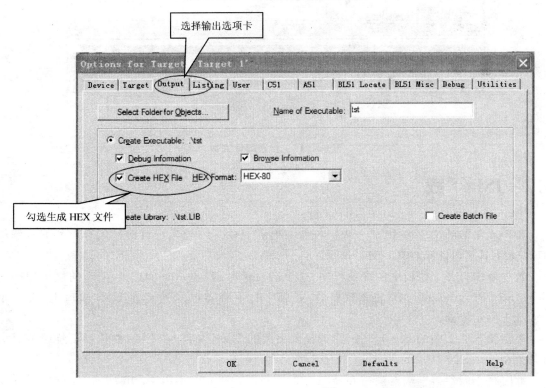

图 6.1.12　Output 配置

第 6 章 Keil μVision4 集成开发环境和 ISP 下载

据提示信息可以对代码进行检错,然后再重新编译即可,直到没有错误。如图 6.1.13 所示,程序编译通过,虽然有警告提示,但不影响程序的运行,是因为部分函数没有调用而引发的警告。

图 6.1.13 程序编译完成

6.2 ISP 下载

按照 6.1 节的步骤操作,将工程进行编译生成 HEX 文件后,就可以使用专用的下载器将 HEX 文件下载到目标 CPU 上运行。

首先双击打开上位机上的烧录程序,这里的目标 CPU 是 nRF24LE1,配套的烧录程序是 nRF24Lxx Programmer。双击该软件后,如果 ISP 下载器已经连接在 USB 端口上,将显示图 6.2.1 所示界面。

单击图 6.2.1 中的 Browse 按钮,选择要下载的 HEX 文件,在 Chip 栏选择要烧录的目标芯片类型。

此时将下载器另一端连接在 nRF24LE1 下载端口上,单击 Program 按钮,进度条就开始显示下载的程序量,完成下载后,烧录软件界面如图 6.2.2 所示,显示程序下载成功。

第 6 章 Keil μVision4 集成开发环境和 ISP 下载

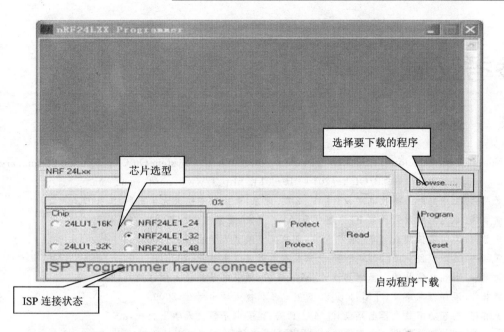

图 6.2.1 ISP 烧录程序界面

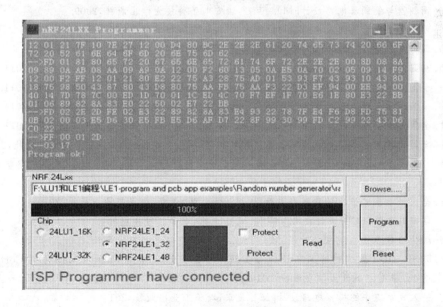

图 6.2.2 程序下载成功

当程序下载完成后，按下 nRF24LE1 电路模块上的复位键，nRF24LE1 即开始运行刚下载进去的程序。

参考文献

[1] Nordic Semiconductor. nRF24LE1 Ultra-low Power Wireless System On-Chip Solution Preliminary Product Specification v1.6[EB/OL]. [2010-10-01]. http://www.nordicsemi.com.

[2] 黄智伟. 无线发射与接收电路设计[M]. 2版. 北京:北京航空航天大学出版社,2007.

[3] 黄智伟. 单片无线数据通信IC原理应用[M]. 北京:北京航空航天大学出版社,2004.

[4] 黄智伟. 无线通信集成电路[M]. 北京:北京航空航天大学出版社,2005.

[5] 黄智伟. 蓝牙硬件电路[M]. 北京:北京航空航天大学出版社,2005.

[6] 黄智伟. 射频小信号放大器电路设计[M]. 西安:西安电子科技大学出版社,2008.

[7] 黄智伟. 混频器电路设计[M]. 西安:西安电子科技大学出版社,2009.

[8] 黄智伟. 射频功率放大器电路设计[M]. 西安:西安电子科技大学出版社,2009.

[9] 黄智伟. 锁相环与频率合成器电路设计[M]. 西安:西安电子科技大学出版社,2008.

[10] 黄智伟. 调制器与解调器电路设计[M]. 西安:西安电子科技大学出版社,2009.

[11] 黄智伟. 单片无线发射与接收电路设计. 西安:西安电子科技大学出版社,2009.

[12] 黄智伟. 通信电子电路[M]. 北京:机械工业出版社,2007.

[13] 黄智伟. 全国大学生电子设计竞赛系统设计[M]. 2版. 北京:北京航空航天大学出版社,2011.

[14] 黄智伟. 全国大学生电子设计竞赛电路设计[M]. 2版. 北京:北京航空航天大学出版社,2011.

[15] 黄智伟. 全国大学生电子设计竞赛技能训练[M]. 2版. 北京:北京航空航天大学出版社,2011.

[16] 黄智伟. 全国大学生电子设计竞赛制作实训[M]. 2版. 北京:北京航空航天大学出版社,2011.

[17] 黄智伟. 全国大学生电子设计竞赛常用电路模块制作[M]. 北京:北京航空航天大学出版社,2011.

[18] 黄智伟. 全国大学生电子设计竞赛ARM嵌入式系统应用设计与实践[M]. 北京:北京航空航天大学出版社,2011.

[19] 黄智伟. 全国大学生电子设计竞赛培训教程[M]. 修订版. 北京:电子工业出版社,2010.

[20] 黄智伟. 印制电路板(PCB)设计技术与实践[M]. 北京:电子工业出版社,2009.

[21] 黄智伟. ARM9嵌入式系统基础教程[M]. 北京:北京航空航天大学出版社,2008.

[22] 黄智伟. 32位ARM微控制器系统设计与实践——基于Luminary Micro LM3S系列CortexM3内核[M]. 北京:北京航空航天大学出版社,2010.

[23] 黄智伟. 嵌入式系统中的模拟电路设计[M]. 北京:电子工业出版社,2010.

[24] 谭晖. nRF无线SoC单片机原理与高级应用[M]. 北京:北京航空航天大学出版社,2009.